Simon David Hirsbrunner
A New Science for Future

Locating Media/Situierte Medien | Volume 26

Editorial

Simon David Hirsbrunner is Senior Researcher at the Human-Centered Computing Research Group of Freie Universität Berlin. He holds a Doctorate in Media Ethnography (Siegen), a Master degree in European Media Studies (Potsdam), and a Diploma in International Relations (Geneva).

Simon David Hirsbrunner

A New Science for Future

Climate Impact Modeling and the Quest for Digital Openness

[transcript]

This book is based on the dissertation »Opening Science for future. The Science & Technology of Climate Impact Prediction«, which was accepted by the Faculty of Philosophy at the University of Siegen in November 2019. The research and publication were supported by the DFG Research Training Group Locating Media at the University of Siegen, Germany. Ethnographic field work for the study was enabled through a visiting fellowship at the Potsdam Institute for Climate Impact Research (PIK). The author would like to thank these institutions and the supervisors, Tristan Thielmann and Cornelius Schubert, for their valuable support.

Bibliographic information published by the Deutsche Nationalbibliothek
The Deutsche Nationalbibliothek lists this publication in the Deutsche Nationalbibliografie; detailed bibliographic data are available in the Internet at http://dnb.d-nb.de

First published in 2021 by transcript Verlag, Bielefeld

transcript Verlag | Hermannstraße 26 | D-33602 Bielefeld | live@transcript-verlag.de

Cover layout: Maria Arndt, Bielefeld
Cover illustration: © Paul Heinicker
Proofread by Philip Saunders
Printed by Majuskel Medienproduktion GmbH, Wetzlar
Print-ISBN 978-3-8376-5265-9
PDF-ISBN 978-3-8394-5265-3
https://doi.org/10.14361/9783839452653
ISSN of serie: 2703-0210
eISSN of serie: 2703-0229

Printed on permanent acid-free text paper.

Table of contents

For Betti and Julius,
who always brought me back
from the future to the present.

Introduction

Unite behind the science.
Greta Thunberg[1]

September 20, 2019. At the very moment I am writing these lines, millions of people are gathering worldwide for a climate strike to pressure governments to take action on climate change. Counting participants distributed around the globe, it is one of the biggest protest demonstrations in human history. The event is closely linked to the youth movement #fridaysforfuture and its lead figure Greta Thunberg, who has haunted selected leaders of Western democracies since the hot summer of 2018. Climate change now sits at our kitchen table, stares at us from our social media timeline and makes our political representatives tremble. Earlier, concern about global warming was linked essentially to certain jobs (environmental sciences, non-governmental organizations [NGOs], journalism), political views (leftist, progressive, ecologist), and ethical considerations (religiosity, protection of God's creation). Now, by contrast, it takes considerable effort *not* to be confronted with the climate issue. Whether one likes it or not, climate change requires active positioning within our daily life, our personal networks and our (analog and digital) filter bubbles. People have tended to take the future with climate change into account within their daily lives only since very recently but now with striking force.

1 https://youtu.be/bz8jSJAkFRM, retrieved on April 3, 2019.

What were the reasons for this changing perception of the urgency of climate change? How did climate change evolve from a scientific fact into a global matter of social, economic and political concern? Attempts at explaining these questions fill many books. Perhaps the new scientific insights about the severe impacts of climate change have convinced the public of its urgency; or humans just needed time to make sense of climate change and figure out ways of dealing with it in a concerted manner; or scientists and activists have learned how to communicate climate change better and beat the argumentation of climate skeptics and deniers. Probably all these explanations have their share of truth. It is, however, undebated that global warming requires new scientific ways of thinking about the world, its past, present and futures.

Figure 1: "Unite behind the Science," tweet by Greta Thunberg following her speech at the French National Assembly on July 23, 2019. Source: Twitter[2]

2 https://twitter.com/gretathunberg/status/1153693427487387648, retrieved on September 3, 2019.

As a matter of fact, there has arguably never been an issue of global concern so intimately linked to the sciences. This relationship is reflected in Greta's call to "unite behind the science" and the central role scientists play in mediating, negotiating, regulating and governing the issue at various levels. Equally important but more controversial is the role of technology in climate change. On the one hand, climate change has literally been caused by technology, namely the massive release of carbon dioxide (CO_2) and other greenhouse gases (GHGs) by multiple industries since the middle of the 19th century. On the other hand, technology also plays a central role in the mitigation of climate change, be it renewable energies, energy efficiency measures or more recent and controversial approaches subsumed under the term of negative emissions.

In this book, however, the focus lies on another technology intimately linked to the climate issue: Computer models. As media scholar Julie Doyle puts it, "climate change has been reliant upon science and technology for its detection, and for predicting the various scenarios of its future development and impacts" (Doyle 2011: 16). Computer modeling has become the fundamental organizing principle for the global epistemic community surrounding the climate change issue (Edwards 2001: 34; Sundberg 2007: 473). This crucial role of computer models and simulations in climate research have also made them a recurrent theme in the social sciences and humanities. Scholars have discussed epistemic and representational issues widely and characterized the way simulations represent the world or aspects of it (Gramelsberger 2008b; Pias 2008; Winsberg 2010). Other works discussed the historical development of climate modeling technology and infrastructure (Edwards 2010) and considered it a lead discipline for the dawning age of simulation-driven science (Gramelsberger 2008a: 105). There has also been considerable academic work addressing the relationships between model-driven climate research and climate policy (Gramelsberger/Feichter 2011), with the exemplary role of the Intergovernmental Panel on Climate Change (IPCC) in this regard (Hulme/Mahony

2010). Scholars have discussed the transformation of climate models from heuristic tools into political instruments (Heymann/Hundebøl 2017) and the sociotechnical frictions created in the course of downscaling global models to the regional level (Mahony 2017; Mahony/Hulme 2012).

Another line of research has specifically considered the role of visual media in the translation of scientific knowledge emanating from computer simulations. Notably, diagrams and maps are popular devices to communicate insights from computer experiments to the broader public (Doyle 2009 2011; Manzo 2009 2010; Schneider 2012, 2017). Climate images gain political status in the process of socialization: "[...] climate science is the paradigmatic field in which images take on a role as political agents" (Schneider 2012)

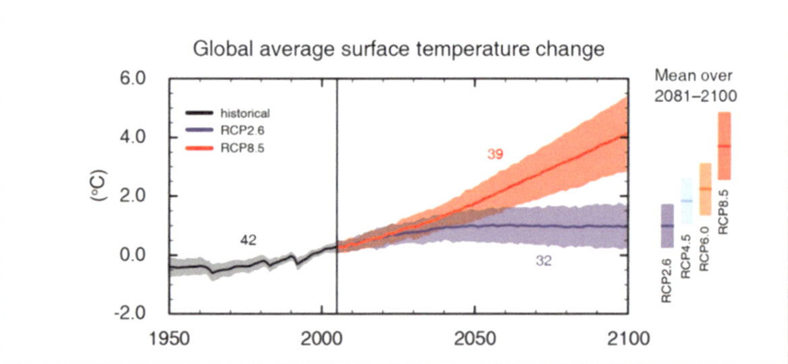

Figure 2: possibility space of the future, drawn as Representative Concentration Pathways (RCPs). Source: IPCC 2013, 21

This is especially true for the visualizations featured in the IPCC assessment reports, notably the "hockey stick" (Montford 2010; Walsh 2014), the "burning worlds" (Schneider 2012, 2016, 2017) and the "burning embers" (Mahony/Hulme 2012). Figure 2 featured in the fifth assessment report of the IPCC (2013), for example, represents the numeric results of comprehensive computer simulations. Nevertheless, they are strategically framed and publicly understood as choices between two pathways of the future – a detrimental scenario (red), with severe warming of the global atmosphere, and a more sustainable scenario (blue), which requires a massive transformation of our ways to produce, consume and live.

As Birgit Schneider has highlighted, these graphs have become mobile across contexts and social worlds. As *immutable mobiles* with considerable semantic flexibility (Latour 1988), they offer various interpretations and fulfill many functions for multiple actors (Schneider 2012).

Mobilization and stabilization

Drawing on this idea of graphs and maps as immutable mobiles for climate change mediation, research for this book started with an ethnographic study of scientific modeling work at the Potsdam Institute for Climate Impact Research (PIK). The PIK is one of the major institutes running comprehensive computer simulations calculating scenarios of the future with climate change. In my interviews, observations and interventions at the institute, I learned a lot about the mobilities and immobilities of visual artefacts within scientific modeling practice. However, I also discovered that images are not by far the only artefacts in climate modeling practice that began to travel. As a matter of fact, many elements in scientific modeling practice have become increasingly mobile, fluid and permeable across social worlds and application contexts. Accordingly, my study also increasingly turned into a multi-sited ethnography (Marcus 1995), involving equally analog, digital and hybrid fields and places of investigation.

In the book, I will document and discuss mobilization and stabilization within scientific modeling practices, which are both linked to new role models in research ('open science') and to the imbrication of modeling practice within digital, networked infrastructures. I argue that these new sociotechnical constellations change the way scientific modeling and climate prediction works considerably. Moreover, these transformations of scientific modeling practice may also provide some explanations for the way climate change has currently 'kicked in' within current debates and various publics.

Technographic ordering of elements

Methodologically, this investigation started with the belief that more qualitative research is needed to understand the practices, artefacts and infrastructures of contemporary data collection, processing, analysis, representation and dissemination. Rather than testing the theories available, the aim of this study was to root theoretical considerations within empirical observations of practice, material and symbolic representation. Barney Glaser and Anselm Strauss have formalized this position in their conceptualization of *Grounded Theory* thinking, which should enable the systematic discovery of theory from data:

> We believe that the discovery of theory from data – which we call *grounded theory* – is a major task confronting sociology today, for, as we shall try to show, such a theory fits empirical situations, and is understandable to sociologists and layman alike. Most important, it works provides us with relevant predictions, explanations, interpretations and applications. (1999: 1)

While generally agreeing on the direction of this verdict, sociologist Werner Rammert has highlighted that there is no such thing as theory-free empiricism in science. Indeed, empirical observations are always focused, put into perspective and mediated. For Rammert, the task is then to document and reflect thoroughly on the selection processes in play, the points of view taken, and the optics or instruments in use (2007: 16). The methodological considerations around *technography* provided some directions and terminological ordering for this study. Werner Rammert and Cornelius Schubert argue against meta-narratives about the ontology of 'the technical' and 'technology' in philosophical and sociological theory in their conceptualization of the approach

(2006). According to them, such meta-narratives may be interesting from the science-historical perspective but fail to grasp the practical discourse with technology, as it is conducted by laboratory researchers, inventors, engineers, workers, entrepreneurs and users (ibid.: 12). This oscillates with my field, where (digital) technology was virtually omnipresent but did not provide a useful category to be 'observed.' General characterizations of 'simulation modeling technology' were not helpful for understanding how actors actually construct computer models, carry out simulations and deal with predictions of the future with climate change. Against this backdrop, technography as a micro-sociological approach aims at investigating the "practical production and installation of techno-social orders in strategically relevant situations, in view of discovering exemplary practices and mechanisms for the development of new institutions and global regimes" (ibid.: 13, translated by the author). Technography may begin with interactions among humans and interactivities with objects and explores patterns of hybrid micro-orders. In conflict and coalition with others, these micro-orders may lead into powerful macro-constellations. Rammert characterizes three types of technology-related practices, namely, *making technology*, *using technology* and *participation of technology* (2007: 4). However, in the case of technological practice in simulation modeling, the boundaries between these categories are blurry or sometimes nonexistent. The technological devices distributed described within this study rarely have distinct 'producers' or 'users' but are mutually constructed and operationalized ('used') by communities of 'contributors.' Then again, such contribution cannot be understood independently from technological participation, which is increasingly automated within digital platforms and infrastructures. In the context of this study, a general categorization of elements regarding the categories of humans, things and symbols has been conducive for identifying interactional relationships.[3] While connections are always drawn between different 'families' of elements, the chapters of the book each have a particular focus – Chapter I explores physical elements, such as architecture, telescopes, computers and forests, Chapter II focuses on the practices of human scientists, chapter III on visual inscriptions, chapter IV on models and software, chapter V on numerical data, and chapter VI on digital platforms, which built relationships between multiple elements.

Technographic interventions

In terms of its methodology and procedural ways forward, technography draws mainly on ethnographic methods of observation, inscription and description. Going beyond classic field studies, these methods include approaches such as videography, webnography and interactivity experiments (Rammert/Schubert 2006: 13f). Compared to these approaches, the present study particularly highlights the inventive capacities of theories, methods and artefacts. In combination, they can be seen as heuristic devices that trigger new questions, draw out new aspects and help to co-create our research fields. Lury and Wakeford (2014) have reiterated the relevance of method to the empirical investigation of the contemporary and suggested the development and operationalization of *inventive methods* to investigate the happening of the social. For them,

> an inventive method addresses a specific problem, and is adapted in use in relation to that specificity; its use may be repeated, but the method is always oriented to making a difference. (Lury/Wakeford 2014: 11)

The authors stress that inventiveness is not to be equated to new. Inventive methods may include well established devices, such as

3 Werner Rammert puts it similarly as "wetware, hardware, and software" (2007: 19).

experiments, patterns or populations, marginal(ized) ones, such as anecdotes, screens or speculations, or relatively new ones, such as the probe or phrase: "What unites them, however, is that they are methods or means by which the social world is not only investigated, but may also be engaged" (ibid.: 6). In so doing, they open up the question of how methods contribute to the framing of change, understood "not only as complex, contradictory and uncertain, but also as everyday, routine and ongoing" (ibid.). Lury and Wakeford are drawn to the term *device,* a word with multiple everyday meanings (object, method, bomb), to conceptualize and operationalize inventive methods further.

> [...] the notion of the device not only admits that object and methods are mutually constitutive, but also acknowledges that it is their relation that forces us to confront the new. (ibid.: 8)

For Lury and Wakeford, devices never operate in isolation but in relation to an apparatus and complex ensemble of practices. This relational and situational quality is a source of permanent destabilization, which prevents the device from becoming "a mere tool, which could be used always and everywhere in the same way" (ibid.: 9). This embedding in the apparatus also highlights the ability of devices to be powerful agents that not only represent reality but co-create it.

I engaged in a number of activities in my fieldwork that oscillate with the idea of inventive methods and devices. This includes a methodological operationalization of maps, interventions such as workshops, and interdisciplinary collaboration. Considering that these devices are embedded in the technographic interventions, they are discussed in the course of the chapters.

Plan of the book

The six chapters of the book are structured around particular elements and field sites. The different scales and ontologies of these sites have evoked different questions, foci, and research tactics. Understanding theory, method, and research phenomena as interrelated categories, which are mutually elaborated within the research process (Bender/Zillinger 2015: XI), conceptual approaches are discussed within the chapters in a dialogue with the empirical material. Accordingly, the chapters aim equally at characterizing distinctive constellations of elements relevant to scientific practice in climate impact research, but also discuss the situated methodological tactics and devices that have been used for investigation.

Chapter I describes the making and consolidation of Telegrafenberg as a place to do spatial, Earth and climate research. In the 19th century, a forested hill near the Prussian residential city of Potsdam was increasingly connected with telegraphic communication networks and put on the map of the German Kaiserreich. In the 1860s, Wilhelm Foerster and other astronomers proposed transforming the hill into a research infrastructure for a prospective new scientific discipline – astrophysics. Potsdam appeared to be a perfect location for this endeavor, due to its position within proximity but also a healthy distance from the Prussian capital Berlin. Building on the work of Susan Star, Karen Ruhleder and Geoffrey Bowker (Bowker 2005; Star 1999; Star/Ruhleder 1996), I will address these aspects of proximity and distance as essential qualities for the making of new infrastructures for technoscientific innovation. This involves diverse forms of infrastructural elements, including physical particles, architectures, rail tracks and telegraph masts, telescopes and spectrometers, scientists and machinists, scientific disciplines and governmental entities. At the beginning of the 20th century, Telegrafenberg had become a veritable place to do astrophysics, meteorology and geodesy. An identification of the elements and their relationships of this achievement is probed through a temporal layering of spatialities (i.e. infrastructure maps) enabling infrastructural inversion (Bowker 2005). Building on Geoffrey Bowker's (2015) comments on the temporality of infrastructures, I also challenge linear representations of time within infrastructural analysis and highlight parallel and repetitive (re)constructions of pasts, presents and futures. The astonishing monument of the Einstein Tower, for example, provides an opportunity to reflect on (dis-)connections from traditions and the promise of futures in architecture and infrastructure. In this context, I also challenge the idea of a master narrative in infrastructure (Star 1999) and propose instead to focus on heterogeneous promises, functionalities and representations that infrastructure provides for different people at different times. Finally, I introduce the newest contributions to the science park and the PIK in particular.

Chapter II is a collage of the idioculture (Fine 1979, 2007) at the PIK. I discuss Hans Joachim Schellnhuber's conceptualizations of Earth System Science (Schellnhuber 1998) and geo-cybernetics (Schellnhuber/Kropp 1998) as the conceptual lines for work at the institute. I propose the entanglement of different dimensions that are specific to the idioculture at the PIK. Firstly, the integration of science for sustainability opens a Pandora's Box multiplying the number of issues and perspectives to be taken into account in research endeavors at the institute. Research at the PIK goes well beyond the calculation of GHGs or temperature increases. It embraces a cybernetic understanding of the world, where everything is connected to everything. Secondly, this multiplication of research objects requires the operationalization of inter- and transdisciplinary collaboration. To be able to make statements about everything, one has to hire experts of heterogeneous scientific disciplines and bring them together to collaborate with each other. Thirdly, the heterogeneity of the scientific disciplines participating in this project has to be stabilized through the use of common scientific methods, namely, those of simulation modeling and computational science. Computer models, code and digital infrastructure are omnipresent elements in the scientific practices at the PIK. Fourthly, I will argue that the PIK should not only be grasped as a center of calculation (Latour 1987), but also as a center of accountability (Rottenburg 2009). Prediction through computers cannot be understood solely as a matter of calculation and epistemic representation but rather as a matter of political representation, translation and accountability. There are different views at the institute how to handle relationships to the world outside the scientific community. As I will argue, these different views are not so much linked to the extent of openness but rather to the question of its timeliness. The principal question is whether one should engage in openness after the stabilization of the facts (i.e. a project cycle) or from the beginning of ongoing experimentation. I introduce the discourse of *open science* and its infrastructural reading as the perspective currently gaining momentum at the PIK. The remaining chapters describe different ways of dealing with openness, mobilizing visualizations (III), software (IV), data (V) and multiple elements (VI).

Chapter III describes the making, operationalization and use of the online geoplatform ClimateImpactsOnline (CIO) as a device for science mediation. Being part of the project team of CIO for one year, I was able to observe and participate in activities of science mediation during the Long Night of the Sciences and educational work within German schools. While the PIK scientists had a clear picture of the 'end users' of the online platform, I argue that the practices within the frame of science communication cannot actually be grasped through categories such as 'users' or 'user interfaces.' In fact, the 'user interface' may have been the one standing in the way of true engagement with the climate issue. Following the further development of the platform and accompanying activities with teachers and pupils, I will show how actors managed to *infrastructure* (Star/Bowker 2006) practices of debate around climate impacts in German schools. This entailed a parallel and continuous configuration of new artifacts and practices. I would argue that the role of the geoplatform in this context was one of an *anchoring device,* enabling open but also channeled debates about climate change and its future.

Chapter IV deals with the mobilization of computer models as a strategy to bring climate change to new territories, communities and infrastructures. Supplementing existing work on the regional downscaling of climate models and its migration to multiple geographic places (Mahony/Hulme 2012; Mahony 2017), I propose to discard representational concerns about traveling computer models and focus instead on technological considerations in model and knowledge mobilization. Discussing the example of the climate impact model CLIMADA, I will show that scientific programming in climate impact research is currently in a state of massive reconfiguration. I will discuss aspects of this reconfiguration, such as Pythonization, coding openness, software packaging, and organization within digital communities and platforms. Furthermore, I introduce the Jupyter Notebook as a methodological device to stabilize such practices of mobilization. Taken together, these configurations can be characterized as *mobile modeling*. The latter, as a practice, generally privileges technologies that optimize performance through the distributiveness of functionalities and elements, while mutually instating devices for overview and control. Mobile modeling is a forward-looking practice aiming at the future amplification of artifacts and knowledge.

Chapter V sheds a light on the related perspective of digital knowledge mobilization, such as open datasets, infrastructures and services. Addressing the making and infrastructuring of a specific open dataset of GDP[4] time series by the PIK scientists (Geiger et al. 2017), I will discuss the conceptualizations of traveling data, information and knowledge in science within existing literature (Latour 1999a; Leonelli 2015; Rheinberger 2011). My observations and infrastructure analysis suggest a shift of attention within computational science from the production of evidence to that of open, linked and reusable datasets. Similar to the production of open software discussed in Chapter IV, open data practices introduce a forward-looking perspective into scientific practice, producing artifacts for prospective reuse. These tendencies of a *datafication of science* are supported by the increasing prominence of the data publication (Costello 2009), which enables researchers to

4 Gross Domestic Product

make 'opening data' accountable within dominant gratification schemes in science (i.e. scientometrics, impact factors).

Chapter VI aims at drawing together several of the aspects discussed so far. It introduces the sea level rise (SLR) mapping project *Surging Seas* as an example of a harbinger of things to come. This includes data-driven predictions of climate change impacts, an increasing permeability of social worlds within the public engagement with science and technology, and the importance of a forward-looking impression within digital platforms and infrastructures. Branching from the concept of fluid technology (De Laet/Mol 2000), I propose a new terminology for a socio-technical understanding and investigation of digital science and technology. Digital technology may seem essentially fluid. However, rather than focusing on fluidity and fluid objects, I propose an attentive shift towards its *viscous elements*. Fluidity and viscosity are different sides of the same coin in fluid dynamics. Viscous liquids may be subject to changes of shape, but they are more persistent than other (more fluid or less viscous) liquids. Viscous elements affect the course of other elements, sometimes even organizing the fluidity of these others. In the context of this study, software libraries, packages and frameworks can be characterized as *viscous elements*. This will be illustrated by the case of the Leaflet Library, which enables the mobilization of interactive maps within the infrastructures and platforms of the web.

The conclusive chapter of the book summarizes the findings of the investigations and discusses permeability and digital openness as new boundary conditions of today's scientific modeling practice.

I. Future infrastructure

> We deconstruct buildings materially
> and semiotically, all the time.
> *Thomas Gieryn (2002)*

Our story begins in the 19th century when Telegrafenberg received its name and gradually became a place to do science.

Ethnographic note, part 1

I walk ten minutes from my apartment in the Berlin city district of Neukölln to Hermannplatz and take the underground U8 to Alexanderplatz. At Alexanderplatz, a major square and traffic hub, I switch from the city transport system BVG to a regional train connecting the German capital with its surrounding federal state of Brandenburg. The train heads west and travels past Berlin's center and sights – Brandenburger Tor, Tiergarten, the government area, Charlottenburg – before entering the rural and forested areas of Brandenburg. I leave the train at Potsdam central station and cross its forecourt, which is usually crowded by international tourists. The city does not only accommodate its 175,000 inhabitants and a vital research community, but is also host to many historical and cultural landmarks of international reputation. An organized touristic visit to Berlin typically includes a one-day trip to Potsdam, where visitors are compensated for the lack of historical buildings in Berlin. Until 1918, Potsdam served as the residence of the Prussian kings and the German Kaiser, equipping the area with a variety of picturesque castles, colorful gardens and spacious parks. Arguably the most impressive architecture, the new Palais commissioned by legendary Prussian King Frederic the Great, now accommodates the University of Potsdam. I briefly walk from the train station through a residential area. I then walk up a small road towards Telegrafenberg, a forested hill about 100 meters high. On the way up, I usually walk besides other wayfarers, all scientists who seem to be absorbed in thinking about their scientific projects, experiments and meetings. The walk up the hill is a boundary time-space, helping to leave the private life in Berlin behind and attuning the mind to scientific work. Arriving at the top, all newcomers to the *Science Park Albert Einstein* need to cross a security barrier with a turnpike and gatehouse, but the contact with the guards is usually limited to an apathetic nodding or mumbled, "Good morning." First-time visitors may stop at a glass vitrine at the wall of the gatehouse, showing a schematic map of the science park, its architecture and infrastructure.

Figure 3: Schematic map of the Science Park Albert Einstein.
Source: Own photo

During my time at Telegrafenberg, I often came back to the vitrine and map, seeking orientation for my investigations. After a while, I substituted the consultation of the vitrine with a paper brochure received from the GFZ press office. In that way, I could walk around and learn while traveling. Later, when I was drawing together the data for analysis, the paper map was again exchanged for a digital version that I found on the web. Such schematic maps of places, institutional structures, machines, models, algorithms and datasets have been valuable data for this study. Some of those maps are also represented in this final text, hopefully providing a means of orientation for the reader.

Ethnographic note, part 2

After passing the security area and some parking lots, the first major building comes into sight on the right: A branch of the *Alfred Wegener Institute (AWI)*, an internationally renowned institute for polar and marine research. Continuing on my way, I pass the German Research Centre for Geosciences (GFZ), a cafeteria and a kindergarten. On the top of the mountain sits a monumental piece of architecture with three cupolas, the Michelson House, today hosting the Potsdam Institute for Climate Impact Research (PIK).

Telegrafenberg had served as an astrophysical observatory for many years before the climate scientists took office at the edifice in 1992. More than that, the Astrophysical Observatory Potsdam (AOP) has been one of the central places for the construction of astrophysics as a scientific field and discipline. Reciprocal manufacture of and experimentation with the telescopic instruments on the hill pioneered the primary method of astrophysics – spectral analysis.

Ethnographic note, part 3

I walk around the architecture, passing through an English garden structure, ending at the Great Refractor, a gigantic optical telescope. On my left, I catch sight of the architectural celebrity on the hill; the Einstein Tower built by

Erich Mendelssohn, regularly environed by groups of international tourists and architectural students. Descending the hill through a beaten trail, I arrive at my final destination - PIK's new office building referred to as the House in the Woods.

Figure 4: Michelson House. The former Astrophysical Observatory, now the headquarters of the Potsdam Institute. Source: Own photo

While I have taken this path to my field site and office hundreds of times, the place has never ceased to impress me with its atmosphere lost in reverie. Working on the hill for some time, one becomes soaked into this spiritual feeling of the place that has been cautiously created, layered and refined for over a century.

The becoming of Telegrafenberg

The Prussian administration under Frederic William III began to construct the first state-run semaphore chain on German ground in 1832. In a time of social unrest and political instability, improving the channels of communication was seen by the Prussian military as a promising mean to establish control over the highly dispersed territories of the Reich: Brandenburg (Berlin, Potsdam) and Rhineland (Cologne,

Koblenz).[5] The semaphore line installed was nearly 600 km long and consisted of sixty-one optical telegraphs, masts about six meters high, passing on encoded information by pivoting shutters or blades (see Fig. 5).

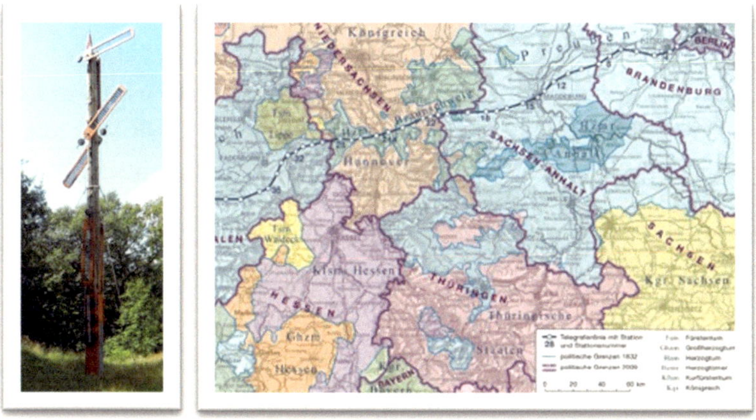

Figure 5: reproduction of the original telegraph on the hill (left) and map of the telegraph line Berlin–Koblenz (right). Source: www.optischertelegraph4.de

The fourth mast was erected on a hill near Potsdam, residential city of the Prussian emperors, and served as its namesake – *Telegrafenberg* (telegraph hill). While the name has survived until today, the semaphore telegraph chain had only a short life as a communication infrastructure and was suspended in 1852, after innovations in the field of information transmission technology, namely electrical telegraphy (Wilderotter, et al. 2005: 88). Nevertheless, the telegraph line had been a major event in the structuration of the Telegrafenberg, drawing physical and symbolic lines between the topographic elevation near Potsdam

5 It is a funny coincidence that the telegraph line connected the locations of my working place (Rhineland), my residential city (Berlin) and my field (Potsdam).

and existing networks and infrastructures of the Prussian administration. It put the location 'on the map,' making it visible, addressable, formable and manageable.

The attribution of the Telegrafenberg as a place to do science can be traced back to the 1860s and 1870s. The ontology of these traces preferences a specific way to tell history – as a history of great men, heroes and inventors. Science and technology studies has always problematized such versions of the 'great man theory,'[6] which limit agency to single, human male, historical figures. This theory translates to diffusionist models for science-society interactions, as illustrated by Bruno Latour:

> Nobody shapes science and technologies except at the beginning, so, in the diffusion model, the only reasonable explanation of novelty lies with the initiators, the first men and women of science. (1987: 134)

We will see later in this study that the contemporary scientific practice is incompatible with such diffusionist views of science-society interactions. However, we also have to consider that historic traces of science are often configured in a way that drives and limits certain interpretations. The primary documents from the 19th century are basically accounts by great men telling the history of other great men, thereby mutually amplifying their relevance in the succession of events. To a certain degree, the present analysis will have to reproduce this narrative structure. We will later give more weight to additional resources that help to disrupt these dominant perspectives on the history of the hill and science park.

---

6 Versions of the 'great man theory' have long been dominant in historical analysis. An example of a theoretical conceptualization is Thomas Carlyle's *On Heroes, Hero-Worship, and the Heroic in History* (1993).

Imagining astrophysics

It was astronomer Wilhelm Foerster, who firs[7] formulated ideas for an institutionalization of astrophysics and construction of an interdisciplinary observatory in Potsdam. Foerster had been the director of the Berlin observatory, which later was named after him (Wilhelm-Foerster-Sternwarte in Schöneberg, Berlin). In his memorandum *Denkschrift betreffend die Errichtung einer Sonnenwarte* of 1871 (Hermann 1975), he conceptualized astrophysics as a new scientific field and argued that this new field of study would require a dedicated place and infrastructure to take shape. He brought up a number of arguments that should convince the Prussian administration to support the construction of the novel facility. First of all, he mentioned a number of scientific discoveries in solar research that could serve as a basis for the scientific practice in astrophysics. As a matter of fact, Prussian scientists had been key in advancing solar research, including the discovery of the sunspot activity cycles (Samuel Heinrich Schwabe), sunspot positions and their rotation (Gustav Spörer), spectral analysis techniques (Gustav Kirchhoff), as well as new insights into aspects of thermodynamics (GFZ 2017: 43; Herrmann 1975: 246). These discoveries changed the way scientists had seen the sun and other stellar phenomena. Previously, the sun had basically be seen as the origin of a tremendous mass attraction and source of powerful light and heat effects. By contrast, people were not so interested in the temporal changes in these dynamics. However, some newer studies revealed the periodicity and magnitude of fluctuations, awaking a new interest in the consequences of these changes for other objects and dynamics within the solar system and universe (Herrmann 1975: 247). Pointing to these new scientific matters of concern, Foerster argues: “Therefore, it should be clear that the

7 Foerster’s proposal builds on ideas formulated by school teacher and scientist Gustav Spörer, who had brought the issue of institutionalization to the attention of the Astronomical Society.

construction of a solar telescope of this kind would be a scientific act of eminence" (ibid.: 250, translated by the author). According to the memorandum, further advances in solar research would only be possible by combining them with the measurements from other fields, such as meteorology and magnetism research. The observations could support nonscientific actors, such as miners and field measurers, in their daily work (ibid.: 250). In that sense, Foerster was a pioneer of not only interdisciplinary research practice but also co-benefits between science and society. Finally, Foerster evoked the 'scientific arms race' between Prussia and other major geopolitical powers of the time, in this case, Great Britain:

> Finally, it should be mentioned that England already operates something similar to the solar observatory described, in its facilities for solar observations and magnetic observations in Kew near London. However, on the basis of the present plan, our construction will be more effective and comprehensive than the one set up in Kew. (ibid.: 251, translated by the author)

In sum, Foerster's letter mainly bows down to three recommendations: Firstly, Prussia should engage in the construction of a new type of techno-scientific instrument, a solar observatory, in order to keep its pioneering role in astronomic and solar research. Secondly, it would be necessary to restructure the entire Prussian research infrastructure for earth and astronomic research to enable effective astrophysical work and discipline scientists behind the becoming field of study. Thirdly, it would be necessary to draw this new conglomerate of scientific institutions together in a new place, supposedly in Potsdam. Of course, a well-formulated letter alone does not make a new institute, but Foerster definitely showed a talent for science-political argumentation mobilizing allies for his ideas. He also had a feeling for constellations, and the letter certainly included the right words at the right time: Considering Prussia's victory over France in 1871 and its strengthened position in Europe's power structures, the time may have been conducive for an institutional reshuffling and financial windfall celebrating the new patriotism. At least, this is how contemporary figures interpreted and narrated the *Entstehungsgeschichte des Astrophysikalischen Observatoriums* (Genesis of the Astrophysical Observatory) a few years later:

> The history of the origins of the Potsdam Astrophysical Observatory teaches us that a stimulating thought alone is not enough to enable such a formidable institute […], but that a political upswing of the fatherland was necessary to enable the realization of the plan […]. (Hurtig 1890: 4f, translated by the author)

It is true that the consolidation of the German Reich after the wars of 1870 and 1871 enabled the release of considerable financial funds, which were invested in not only trade and commercial matters but also the arts and the sciences. Consequently, a political blessing for the Telegrafenberg project was given immediately after the German victory (ibid.: 4f).

Installing a base for astrophysics

We will now attempt a change of perspective, drawing away from great men and emphasizing the role of material infrastructure in the making of science. If one is able to listen, infrastructures tell intriguing stories, in this case, stories about scientific engagement for over a century. As Susan Leigh Star has highlighted, infrastructures can serve as information-collection devices for investigations into science, technology and society (Star 1999: 387). By cumbersome work, we may be able to bring to the surface the master narrative of infrastructure, which normally operates invisibly and unnoticed, but no less powerfully in the background of human activities; or, as Star describes it, "a single voice that does not problematize diversity," which "speaks unconsciously from the presumed center of things" (ibid.: 384). Employing a term coined by Geoffrey Bowker, the aim is to engage in an "*inversion of infrastructure*" (Bowker 1994), taking a firmly temporal and relational perspective on such structures. Star and Ruhleder (1996) have famously highlighted this aspect: Instead of asking *what* infrastructure is, we should think more about *when* an infrastructure is. With this temporal definition of infrastructure, they challenged the commonsense understanding of infrastructure as a stable material structure that is "just there":

> Common metaphors present infrastructure as a substrate: Something upon which something else "runs" or "operates," such as a system of railroad tracks upon which rail cars run. This image presents an infrastructure as something that is built and maintained, and which then sinks into an invisible background. It is something that is just there, ready-to-hand, completely transparent. (ibid.: 112)

According to the authors, this understanding of infrastructure of a "system" and "thing" is problematic, as it fails to capture relationships between practice and technology properly: The internet may serve as an infrastructure to support communication for some but fails to do so for others (e.g. a blind person). For the plumber, the water system is not a background support infrastructure but the crucial target object in his/her daily work (ibid.). We should think about infrastructure as something that is relational, temporal, built *in situ* rather than as stable structure serving everybody. These aspects are deeply interwoven in the sense that infrastructure merges when a constellation of related elements is reached within a situation, enabling the functioning of another system. This relational argument builds on the work of others, such as Yrjo Engeström (1990), Geoffrey Bowker (1994) and Gregory Bateson (1987). Bateson argued in his *Steps to an Ecology of the Mind* (1987) on a general level that "What can be studied is always a relationship or an infinite regress of relationships. Never a 'thing'" (Bateson quoted in Star/Ruhleder 1996: 112). In the following, we will trace the established relationships between infrastructural elements in the becoming science park on Telegrafenberg. To assist the reader, the text is loosely structured as a linear timeline, beginning in the early 19th century and ending in the present. However, occasional disruptions of these temporal linearities are needed to grasp the essence of the science park as infrastructure.

Adding a place to the map

The Prussian parliament commissioned the construction of the AOP on Telegrafenberg, a forested hill near Potsdam, in its winter session of 1873/1874 (Hurtig 1890). The AOP was not only the first astrophysical research facility in the world, but also the first of Prussia and Germany's non-university scientific institutes, which since then, have incubated a multitude of techno-scientific innovations. The design, planning and oversight of the construction work for becoming AOP was charged to the relatively unknown architect Paul Emanuel Spieker. Why did the Prussian administrators and scientists involved choose Telegrafenberg for this establishment of astrophysics – a remote location for a highly prestigious infrastructure? Why not build in Berlin or another major city of the Reich? As a matter of fact, such a construction within the capital had been planned initially. However, the first idea had then been discarded due to the fast expansion of Berlin, which caused severe problems for astronomical observation. The fast urbanization had already stripped away much of the functionality of the existing observatory built in the 1830s.

> Thus one has turned one's gaze here to the south bank of the Havel, where large forest complexes were in the possession of the state and situated high up. An outstanding place, 'Telegraphenberg,' was found to be particularly suitable. It rises with its highest peak up to 95 meters above the zero position of the Amsterdam level. (Spieker 1879: 1, translated by the author)

According to Spieker, the Telegrafenberg fulfilled all the requirements for the experiments imagined to be carried out in astrophysics:

> The high position provides a free horizon in all possible directions, while the surrounding forest area in the hands of the state keeps away any disturbing settlements. At the same time, the vegetation prevents the heat radiations, which are detrimental to the observations. (ibid.)

The susceptibility of the sensing technologies to disruptions from the outside and the resulting issue of biased data has been a major issue in astronomy similar to other scientific disciplines dependent on measurements. It has long been clear that data bias cannot be achieved by deleting all environmental influences but by carefully choosing or redesigning a 'natural' habitat. This suitability of parklands to serve as a habitat for scientific sensing technologies had also been widely discussed in the sciento-architectural literature of the time (Wilderotter et al. 2005: 89f).

On the other hand, Potsdam was a convincing location through its relative proximity and connectivity to the networks of political power in Berlin: "The location, [...] the proximity of the Berlin-Potsdam Railway, provides a convenient connection to the outside world, and the capital in particular" (Spieker: 1879: 1, translated by the author). As Wilhelm Foerster had already argued in his memorandum of 1971, Potsdam was located at a perfect distance from Berlin; far enough to prevent disturbances from the cacophony of signals and politics but still near enough to ensure an attribution of scientific work to the capital and empire: "It would be advisable to set up the institute not in Berlin itself, but in such close proximity to the capital that its achievements will benefit the scientific reputation of the capital" (Herrmann 1975: 251, translated by the author).

Instrument, representation, and support

The new science compound in Potsdam was originally imagined as a twofold structure, including an astrophysical institute (for the study of the sky) and a telluric institute (for the study of the earth). However, the concept was dropped due to political (fear of exceeding the concentration of power) and organizational concerns (unmanageability of an oversized telluric institute) (Galle 1926: 67; Hurtig: 1890: 6). More pragmatically, the astrophysical institute was soon complemented by the Royal Observatories for Meteorology (1890) and Geodesy (1892). The spatial composition and aesthetic design of the science park aimed at a reconciliation and demarcation between different purposes of architectures: Representation, scientific instrumentation and support of scientific work.

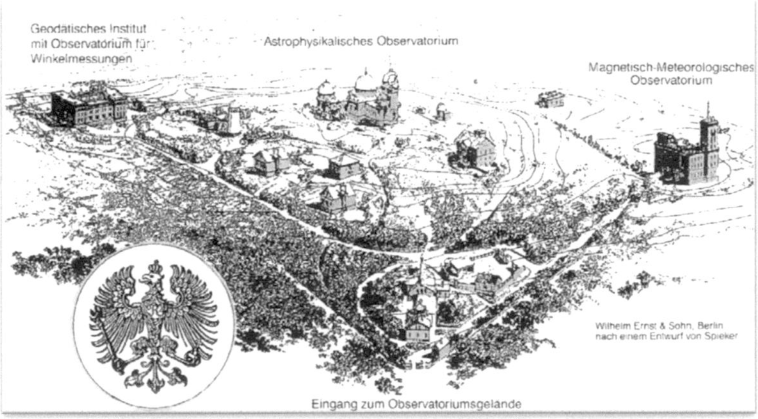

Figure 6: The Royal Observatories of the Telegrafenberg.
Source: Boch (2008)

These aspects of reconciliation and demarcation can be illustrated by an account of the astrophysicist A. Galle from 1926:

> The buildings on the Telegrafenberg site were positioned according to technical requirements. Nevertheless, one has the impression that one is standing in a palace garden with structures arranged pursuing aesthetical considerations. They seem almost like mosques of an oriental city and appear like luminous stones in the frame of Potsdam's city- and landscape. (ibid.: 65, translated by the author)

As an illustration of the time shows (see Fig. 6), the main buildings of the three institutes (geodesy, astrophysics and meteorology) were built almost equidistantly on a straight line from north-west to south-east. Together with the entrance, they form an isosceles triangle. The position of the main buildings, on the top of the hill and along the triangle base, demarcated the equality and congeniality of the three scientific fields and institutes. All other constructions, along the two legs of the triangle at a lower elevation, hosted supporting infrastructures that *enabled* such scientific work but were not seen as being *part* of the scientific processes. Spieker also confirmed this demarcation in his construction report:

> These subordinated installations may have been sufficiently described by now, considering that their characteristics and use are hardly of general interest. In the following, only the more important buildings for scientific purposes will be subject to a deeper discussion [...]. (Spieker 1894: 6, translated by the author)

As a matter of fact, many of these "subordinated installations" were the first ones erected on the hill (Hurtig 1890: 8). These structures served as security (gatehouse), staff accommodation (director's house), maintenance of technology (accommodation for the machinist personnel), provision of alimentation (farmyard) and water (a well system), and electricity (gasworks, generator).[2] Telegrafenberg had been electrified years before the neighboring (major) city of Potsdam (Spieker 1894: 1).

An intermediate position in this demarcation between different infrastructural classes is taken up by scientific technology. Technology at the science park was highly visible (and is still today). The most prominent examples within the 19th century park were optical and photographic telescopes. Considering a focus on astrophysical experimentation, the telescopes of the Potsdam observatories had to be very different from the traditional ones in astronomy. In the latter, telescopes were used to measure the positions of celestial bodies, thereby enabling a continuous and ever more exact and detailed mapping of the sky. By contrast, telescopes serve as light collectors for spectral analysis within astrophysical experimentation. Spectroscopy was the key scientific method in the becoming field of astrophysics. The theoretical fundament of the method was laid throughout the 19th century. Joseph Fraunhofer had already noted dark lines in the spectrum of sunlight at the beginning of the century, but a formalized interpretation of this phenomenon was only given much later by Gustav Kirchhoff and Robert Bunsen. Their development of spectral analysis made it possible to determine the chemical composition and physical state of hot gases and vapors. Scientists were then able to make statements about a star's physical states and processes, its chemical composition and its dynamics by dispersing the light according to wavelengths (GFZ 2017: 46). Potsdam's Telegrafenberg had been a major place to conceptualize, probe and further refine the instrumental settings for these novel experiments. Over time, the setting on the Telegrafenberg enabled groundbreaking experiments, such as those by physicist Albert Abraham Michelson. Michelson carried out his first interferometer experiment in the basement of AOP in 1881, aiming at a scientific proof of ether, a postulated medium for the propagation of light (Michelson 1881). The 'failure' of these experiments (i.e. to identify ether) was one of the theoretical prerequisites for the special theory of relativity described by Albert Einstein in 1905.

Layering infrastructure

Once disruptive for existing structures, science on Telegrafenberg gradually became institutionalized and stabilized. On a material level, the architectural compound designed by Spieker gradually became the infrastructural base for all scientific work on the hill. The functions of the numerous buildings there have changed many times since then, but they are still the formative structure of the entire place. The relationship between the existing structures and arriving newcomers is not always without friction. Be it by submission or rebellious behavior, all subsequent architectures, institutes and scientists had to position themselves regarding the historically accumulated and arranged elements already in place. As Star and Ruhleder highlight,

> Infrastructure does not grow *de nova;* it wrestles with the 'inertia of the installed base' and inherits strengths and limitations from that base. Optical fibers run along old rail-road lines; new systems are designed for backward-compatibility; and failing to account for these constraints may be fatal or distorting to new development processes [...]. (Star/Ruhleder 1996: 113)

As we will see later in this study, these characterizations are equally true for the case of physical infrastructures as for those within the digital realm. In the following, I will discuss some tactics that have been helpful to unravel the 'inertia of the installed base' and the layering of infrastructure at Telegrafenberg. During my investigation of Telegrafenberg as an infrastructure, I have collected a variety of visual representations of the hill's architectures. These include my own photos of buildings, all sorts of maps and plans, drawings, and satellite and drone imagery. The most instructive ones have probably been the schematic maps, such as that shown in Figure 7.

Figure 7: Layers of infrastructure on Telegrafenberg.
Source: http://www.optischertelegraph4.de/telegraphenberg/index.html, retrieved on July 4, 2019.

The maps show different generations of buildings, the original Spieker complex (1874 – 1893), supplements until World War II (1899 – 1928), the era of the cold war and German division (1950 – 1985), and the period from 1990 to the present. Different classes of buildings are highlighted according to their time of formation (map layer, color) and geometric position (auxiliary lines). The maps have been designed by a number of current and former employees of the GFZ, maintaining a website about Prussian telegraphy (www.optischertelegraph4.de) and the history of Telegrafenberg. This work of (what we might call) 'citizen scientists' has been exceptionally helpful as a starting point for my own investigations.

Similar to me, the developers of the 'optischertelegraph4' website engage in activities of infrastructural inversion, surfacing what is normally kept invisible (Bowker 1994; Star 1999). One has to consider that such activities do not only surface but also co-construct certain patterns and perspectives. Accordingly, the maps and descriptions of 'optischertelegraph4' reflect how its developers perceive reality and make sense of the world around them. In this particular case, this view is framed by techniques of professional vision (Goodwin 1994) operationalized in geodesy and the geosciences. The maps of 'optischertelegraph4' helped me to navigate the time-spaces of the science park and to make connections between different elements. As for the 'intertia of the installed base,' all the buildings constructed on Telegrafenberg between 1991 and 2010 pay tribute to the 19th century base by mimicking its geometric forms, materials and colors. The clinker brick facades of the new GFZ and AWI buildings, which were both erected in the late 1990s, are an example of this eclecticism. On the one hand, this meant subordination to a certain tradition of scientific practice that is represented by the Spieker architecture (i.e. the natural sciences, geo- and spatial sciences, physics). On the other hand, it also enabled the new institutes to inherit the reputation of all the scientists, instruments and institutes ever hosted by the science park.

Apart from these aspects of inheritance, the Spieker architectures have been repurposed multiple times in view of changing infrastructural entanglements and challenges. The former house of the director, for example, now serves as a kindergarten and today's 'Café Freundlich' was originally the residence of astronomer Erwin Freundlich (see next paragraph). During my time at PIK, I had the honor of giving a presentation at the 'great cupola,' which once hosted the main telescope of the AOP. Finally, the former Institute for Geodesy has now been repurposed as a central library shared by all organizations on the hill. The library does not only store and make available publications in book form but has become a central node providing digital infrastructure to the institutes of the science park, most notably open data repositories, platforms and services (see chapter V).

Promises of infrastructure

Concentrating on inheritances and disruptions in the time-space of Telegrafenberg as an infrastructure, some elements fairly drop out of the formalistic discipline; these include the Einstein tower, the remains of architecture built during the Cold War era and the House in the Woods. In the following, I will concentrate on the first element – the Einstein Tower.

Figure 8: Einstein Tower. Source: own photo

The history of Telegrafenberg is interwoven with the physical theory of relativity and the person of Albert Einstein for a number of reasons. The material manifestation of the connection to the physicist was the construction of the Einstein tower in 1926 and the experiments undertaken within its solar observatory. But equally, the connection to the famous scientist has been enforced strategically by branding Telegrafenberg as *Science Park Albert Einstein* in 1992. The driving force behind the construction of the Einstein Tower was the mathematician, astronomer and astrophysicist Erwin Finlay Freundlich, who had worked at the astronomic observatory in Berlin and was in regular contact with Albert Einstein. Einstein had repeatedly urged the necessity of proving his theory of relativity by empirical means. He wrote in a letter to Freundlich in 1913: “Theory is not the way forward here” (Wilderotter et al. 2005: 136, translated by the author). Einstein was referring in his letter to the expected eclipse of the sun in 1914 and the opportunity to prove one of the conditions for the general theory of relativity in this context, namely, gravitational lensing; the distribution between a distant light source and an observer that is capable of bending the light from the source as the light travels towards the observer.[8] In order to engage in empirical investigations of the theory of relativity, Einstein managed to organize funding for an expedition and experiments to be carried out by Freundlich. Unfortunately, the First World War rendered the implementation of this scientific project impossible. During the war, a number of scientists engaged in empirical experiments trying to prove or disapprove Einstein’s theory. Potsdam astrophysicist Karl Schwarzschild, for example, tried to prove the relativistic redshift of solar spectral lines by means of a small apparatus installed on the roof of the employee accommodation on Telegrafenberg. However, it appeared that the data available did not dispose the necessary qualities to enable the essential proof of the theories. It became apparent that this objective would only be achievable with an instrument that would first have to be invented and constructed. This ambitious project was tackled in 1920, when architect Felix Mendelsohn was given the opportunity to design the shell for a sun observatory on Telegrafenberg, near the AOP installations. The tower should unite a domed observatory with an underground laboratory. The specifications by Erwin Freundlich determined the geographic location, general elevation and ground plan of the building and Mendelsohn should design and build the appropriate architectural structure to host the instrumental setting. Under these circumstances, Mendelsohn created a building (see Fig. 8) that had been perceived as eccentric and untraditional as the scientific

8 https://en.wikipedia.org/wiki/Gravitational_lens, retrieved on July 4, 2019.

concepts to be approved inside. It brought Mendelsohn great fame but also created a lot of controversy in architectural circles.[9] German publisher and art historian Paul Westheim may have characterized Mendelsohn's working practice and architecture appropriately when he wrote in 1926:

> What architecture is really about, he does not seem to know, but also does not care. If he was more of a proper architect, his construction [...] would have more structural consistency, but probably also much less of the swing, through which he draws attention to himself. He has the grandiose self-confidence that is peculiar to the genius and the dilettante. His technical unorthodoxy may hinder him from being a master builder in the true sense of the word. But it also allows him to manage the construction material in a naïve way, from which many architectural professionals who think of matters of functional design would shy away. (Westheim quoted in Wilderotter et al. 2005: 101, translated by the author)

It is undisputed that Mendelson ignored major traditions of statics and structural engineering. Four years after its finalization, the tower was in severe need of renovation and was characterized as a 'construction error.'[10] The damage was so severe that a first comprehensive renovation had to be carried out as early as 1927/28. In this process, nearly all horizontal components were reinforced by sheet metal and the originally textured outer walls were smoothed (ibid. 2005: 117). Interestingly though, these functional deficiencies did not jeopardize the critical reception of the tower, which today is considered as one of the major style-forming objects for expressionist and organic architecture.

A recurring theme in the critical reception of the tower has always been the relationship between Mendelsohn's architecture, Einstein's

9 See https://c20society.org.uk/botm/einstein-tower-potsdam/, retrieved on July 3, 2019.

10 Neologism by Christine Hoh-Slodczyk, cited in Wilderotter et al. (2005: 117).

theoretical work and the scientific experiments carried out in the tower by astronomer Freundlich and his team. Art historian Fritz Hellwag, for example, declared in 1926 that the architecture of the Einstein tower represents the embodiment of a new age of physics:

> Just as Einstein's discovery represents a sharp cut from previously imaginable research activities, so too has [the Einstein tower's] architect used new construction methods that have hardly anything in common with those practiced earlier. (Hellwag 1926, cited in Wilderotter 2005: 9, translated by the author).

Against the 'newness' of the Einstein tower and theory, cultural and art historian Hans Wilderotter has argued that the tower had not only been erected by quite conventional means, but that it also had some references to older architectural traditions. Equally,

> [...] the new physics had numerous lines of connection to classical physics, as Albert Einstein repeatedly emphasized, and that research at the Einstein Tower would have been unthinkable without the pioneering spectral analyses of the Astrophysical Observatory. (Wilderotter 2005: 10, translated by the author)

As these differing interpretations illustrate, the emphasis on tradition or innovation is also a matter of framing and political choice. Nevertheless and corresponding with the Thomas theorem, the fact that people believe in the reality of a situation, the latter are real in their consequences (Thomas and Thomas, 1928). It is undisputed that the Einstein Tower had been seen as a sign of the beginning of a new era in architecture, physics and beyond.

Once constructed, the Einstein tower enabled the conduction of various experiments on solar spectral analysis. In fact, the installation is still in use today by the successor of AOP, the Leibniz Institute for Astrophysics Potsdam (AIP). On the occasion of the Long Night of the Sciences, a yearly event opening the doors of science to the public, I was able to visit the inside of the Einstein Tower, including its underground laboratory. A current employee of the AIP introduced me to the way the sun observatory produced and hosted data documenting sun activity by means of photogrammetry.

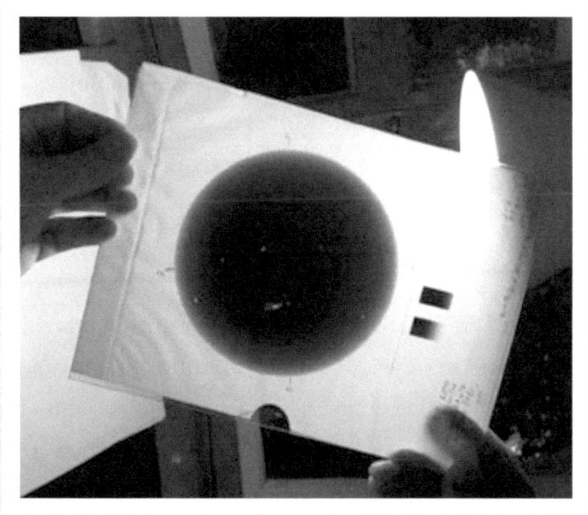

Figure 9: 'Data' in astrophysics. Photo Source: own photo

Figure 8 depicts an example of such 'data.' It shows a negative photographic image of the sun, inscribed as a layer of photo emulsion on a glass plate. As the AIP scientist was highlighting, the data collected and stored since World War II[11] is extremely valuable for contemporary research on sun activity. As a result, the data inscribed on the photogrammetric plates are currently digitalized to make them deployable in state-of-the-art spectroscopic technology and infrastructure.[12]

11 The data collected before World War II have been destroyed as a result of the bombardments of Telegrafenberg by allied forces.

12 Information gathered and photo (Fig. 9) taken during discussions with AIP personnel at the *Long Night of the Sciences* 2019 (June 15). There is more

Considering the interest in the architecture, guided tours had to be offered to the public. The scientists at the Einstein Tower were not only expected to engage in scientific experimentation but to represent these practices on a daily basis. They r became accountable to not only funding agencies and political administrators but an unspecific and unpredictable new category of 'the public.' The task of providing the guided tours was taken up by Harald von Klüber, an employee of scientific director Erwin Freundlich. Von Klüber soon had to learn that people were not interested in the techno-scientific nitty-gritty of astrophysics but absorbed by the monumental architecture on the hill. To deal with this situation, von Klüber experimented with analogy: He did not only relate the theory of relativity to the architecture of the Einstein tower, but put these into perspective with older structures on the hill and more traditional theories in physics (ibid.: 9). Within this storyline, the once revolutionary AOP had become a symbol of the 'old,' while the Einstein tower represented the 'new,' the innovative. This relational set between material structures, architectural patterns and physical theorems had a very improvised character. Von Klüber did not have at his disposal an expertise about architectural design and its history but a lot of knowledge about 'old' and 'new' physics. Nevertheless, his improvised narrative strategy had been quite popular on the hill and was echoed by cultural commentators of the time. And Albert Einstein? It is said that the physicist was conservative in his architectural taste and could not really connect to Mendelsohn's tower. After having been shown through the building by Mendelsohn, he gave a one-word review by whispering into the architect's ear: "Organic."[13]

---

information on the digitalization project APPLAUSE via https://www.aip.de/de/aktuelles/scientific-highlights/historische-stern-daten-digital-verfuegbar, retrieved on June 3, 2019.

13 https://c20society.org.uk/botm/einstein-tower-potsdam/, retrieved on April 2, 2019.

Division and reunion

The Einstein tower and many other structures on Telegrafenberg were severely damaged during World War II. Many scientific facilities, instruments and data collections were destroyed and it took the infrastructure considerable time to recover. Later, during the time of the German division, the German Democratic Republic (GDR) and cold war, the institutes on the hill were reorganized and merged into the Zentralinstitut für Physik der Erde (ZIPE; the Central Institute for the Physics of the Earth). *Inter alia*, ZIPE included the former Institute for Geodynamics (Jena), the Geodetic Institute (Potsdam), the Geomagnetic Institute (Potsdam and Niemegk), the Geotectonic Institute (Berlin) and the GDR working group for extraterritorial geodetic and geophysical research in Potsdam (Kautzleben 1999: 34). This fusion was accompanied by disciplinary research restructuring, leading to the establishment and emancipation of *cosmic physics* as a new field of research, and the strengthening and emancipation of geo- and astrophysical sciences within the GDR. Cosmic physics included the subdomains of astrophysics, earth physics, solar-terrestrial physics, oceanography and geography (ibid.: 35). A number of new buildings were erected on Telegrafenberg during the communist reign in East Germany. As was often the case in the GDR, the constructions purposely broke with the architectural traditions of the Kaiserreich and emphasized other aspects, such as functionality and social utility. As the historiographers of optischertelegraph4.de note on their website, the GDR buildings "[…] were not always erected with a noticeable geometric connection to the existing buildings."[14] Based on our observations about the Einstein Tower, we could go further than that and argue that the GDR spared no effort to make clear that its new facilities broke with every single aspect of the hill's traditions – in a geometric, aesthetic and

14 http://www.optischertelegraph4.de/telegraphenberg/index.html, retrieved on April 2, 2019.

political sense. As we have seen before, the embeddedness of infrastructures (Star/Ruhleder 1996: 113) involves more than physical structures (e.g. optical cables along old railway tracks) and includes social arrangements and technologies.

The story repeated itself after the fall of the Berlin wall, when most of the GDR constructions were torn down (illustrated by the maps in Fig. 6). The reunited Germany had a strong desire to remove the traces and reminders of its painful division. By contrast, it began to restore the architectural remains of its 19th century scientific grandeur. All employees of ZIPE were dismissed in 1991 and a new restructuring on the hill took place. Some existing structures and some of the ZIPE personnel were taken over by a newly founded institute for geosciences, now operating under the name of the *Helmholtz Centre Potsdam – GFZ.* A comprehensive architectural compound was built to host this newly established hub for the geosciences in Germany as a replacement for the GDR buildings demolished. On the top of the hill, within the former headquarters of the Astrophysical Observatory, another institute became part of the science park: The Potsdam Institute.

The PIK is now a major global player in the fight for climate change, but the institute started small in 1992. The German Federal government decided to found a climate institute prior to the environmental summit in Rio de Janeiro (1992) to show its commitment and demonstrate its leading role in matters regarding sustainability. Hans Joachim Schellnhuber, then Professor for Theoretical Physics at Oldenburg University, saw an opportunity and offered himself and a concept (see chapter II) for the institute. The research center started with about 30 employees, mostly physicists, in Berlin in 1991. Two-thirds of the employees had been taken over from the Academy for the Sciences of the recently collapsed GDR. Manfred Stock, one of the then employees remembers the first two working days:

> [...] we drove to Normannenstrasse in Berlin and entered a building within the former Stasi[15] headquarters. These were our first offices. Cameras were staring at us from everywhere. (Hoffmann 2017, translated by the author)

However, the researchers could not stay there for long and the search for a new location for the institute building began. The result was Telegrafenberg in Potsdam. Stock, who was deeply involved in the search, remembers: "We first moved into a container construction – temporarily, as we were told. But the temporary measure then had to hold until 2001" (ibid.: 2017). Eventually, the PIK researchers were allowed to move into the prestigious Spieker architectures distributed on the hill, such as the former Astrophysical Observatory (now 'Michelson House') and the Meteorological Observatory (now 'Süring House'). This set was supplemented by a new construction built specifically for the PIK, the 'House in the Woods,' in 2015.

Heterogenous temporalities of infrastructure

In ethnography, the outcome of analysis may resemble more of a collage than a traditional master narrative with a single voice. As sociologist Herbert Kalthoff has argued, a collage is not about theoretical saturation but about the mobilization of different relevances. For him, the methods and elaborated research results represent contexts for each other which are enriched by contradictions and frictions. They do not have to be brought into agreement:

> Such a collage documents, firstly, the constructiveness of the research methods, secondly, the tension between the individual elements and, thirdly, the aesthetics created by the materiality and arrangement of elements. (Kalthoff 2010: 363, translated by the author)

15 *Ministerium für Staatssicherheit*, the secret police agency of the GDR.

Our collage of the Einstein Tower and its entanglements has implications regarding the characterization of infrastructure: It highlights the crucial aspect of future temporalities, imaginaries and promises for the being of infrastructure. The related structures on Telegrafenberg do not only represent a great past and are functional in the present, they also make bold promises about the future. As Nikhil Anand, Akhil Gupta and Hannah Appel have highlighted in their anthology on *The Promise of Infrastructure*:

> Material infrastructures, including roads and water pipes, electricity lines and ports, oil pipelines and sewage systems, are dense social, material, aesthetic, and political formations that are critical both to differentiated experiences of everyday life and to expectations of the future. They have long promised modernity, development, progress, and freedom to people all over the world.
> (Anand et al. 2018: 11)

Every infrastructure makes such promises about the future: A library promises to make books (i.e. knowledge) available for anyone, the internet promises to connect all human beings on the earth, and highways can potentially take you anywhere. The differences between infrastructures may often bow down to different promises they make about the future and how explicit they are about their future imaginary. As a matter of fact, I prefer the term 'promise' to related terms of the 'imaginary' (Jasanoff/Kim 2015) and the 'master narrative' (Star 1999). Infrastructures do not always speak with one single voice or construct a homogenous imaginary. By contrast, they may well promise different things to different people at different times. The Einstein Tower as a research infrastructure is a good example of such a collage of future promises, but so are other buildings on the hill. I will introduce PIK's 'House in the Woods' and its promises of a future of deep sustainability in chapter II.

II. Future work

After entering the *Albert Einstein Science Park*, I pass by several buildings of the GFZ, the canteen and kindergarten shared by all institutes. On the top of the hill, I walk by the imposing main building of the Potsdam Institute and step into the woods again. After about 50 meters, one can differentiate between the fabric of the forest and another structure built by humans - the newest building on the Telegrafenberg, adequately referred to as the *House in the Woods*.

Figure 10: The House in the Woods. Source: Own photo

At the time of my first arrival, the building was in its last stage of finalization and the smell of the recently painted walls was still filling the air. Considering that the house was not completely occupied by PIK staff, I generously received my own office on the second floor of the 'Earth Wing.' The cloverleaf form, the wooden façade, the tortuous inside and the illustrious designations (e.g. 'Heaven Wing,' meeting room 'Africa') evoke a spiritual feeling between anthroposophist and Silicon Valley aesthetics. The building does not only fit well into its environment - it is a materialized model for the way we ought to build, work and live in the future. This representation of a sustainable future becomes visible prior to entering the building or even looking at the façade. The front yard of A56 is marked by an immense bicycle parking space and construction, complemented by a small parking lot for electric vehicles.

Figure 11: The front of A56 with the parking area for bikes (left) and electric cars (right). Source: Own photo

On my first day at the PIK, scientist Tim Neitzel kindly welcomed me at the House in the Woods. The PIK scientists had just moved from one of the remaining constructions of the GDR era to their brand-new office building, the latest architectural addition to the science park. I received my own office, which happened to be the 'model office.' Furnished by an architectural agency, it simulates the typical equipment of a scientist. Accordingly, I found myself in the strange position of a model researcher in a model office investigating the culture of modeling work.

About climate change/research

Characterizations of climate change fill countless books and take many forms. Climate scientists usually understand climate as 'averaged weather,' as can be illustrated by the following definition by the World Meteorological Organization:

> Climate in a narrow sense is usually defined as the 'average weather,' or more rigorously, as the statistical description in terms of the mean and variability of relevant quantities over a period of time ranging from months to thousands or millions of years. The classical period is 30 years, as defined by the World Meteorological Organization (WMO). These quantities are most often surface variables such as temperature, precipitation, and wind. Climate in a wider sense is the state, including a statistical description, of the climate system. [16]

Such averaged weather patterns may change due to natural variability ("climate variability") or – increasingly the case – because of human activities ("climate change"):

> 'Climate change' means a change of climate which is attributed directly or indirectly to human activity that alters the composition of the global atmosphere and which is in addition to natural climate variability observed over comparable time periods. (United Nations 1992)

---

16 WMO website: http://www.wmo.int/pages/prog/wcp/ccl/faq/faq_doc_en.html, retrieved on April 23, 2019.

This characterization by the United Nations Framework Convention on Climate Change (UNFCCC) from the year 1992 points us to another aspect of climate change as a concept – its intrinsically political nature. This consideration of the political consequences of human-made climate change (a pleonasm, so to say) had already been established as early as 1979, when the Charney report (Charney et al. 1979) provided the first comprehensive assessment of global climate change triggered by increasing carbon dioxide (CO_2) emissions. As Gabriele Gramelsberger and Johann Feichter have shown, the report transformed climate change into a public policy issue, "interlinking climate science and politics by establishing a growing number of international research programs, conferences, WGs, intergovernmental panels, and committees" (2011: 2). Since then, a multitude of activities have been conducted and various institutional organisms established to smoothen the interface between science and policy, most prominently represented by the establishment of the IPCC in 1988. The dominant way conceptualizing science-policy-society interfaces through the vehicle of the IPCC has been described and criticized by a variety of actors, including authors in science and technology studies (Hulme/Mahony 2010; Shackley/Skodvin 1995; Yearley 2009). Among other things, these authors called attention to the strong disciplinary bias of the IPCC towards the natural sciences,[17] and the virtual absence of accounts from interpretative social sciences. Accordingly, Steven Yearley has argued that

> [...] the focus of analyses of the debate over climate change has – understandably– been fixed on the natural scientific aspects of the issue as represented in models of the climate, oceans and atmosphere operated by scientists associated with the IPCC (Intergovernmental Panel on Climate Change) and others. [...] This orientation has led to a neglect of the importance of the ways that economic and social scientific aspects of global warming have

17 The only social science discipline marginally represented within IPCC reports has long been (macro-)economics.

> entered into the business of forecasting, understanding and trying to manage the changing climate. I thus propose to set out and exemplify the case for refocusing attention on to the social science aspects of climate change. (Yearley 2009: 390)

This critique widely shared within the social sciences gave way to numerous studies on the social and cultural aspects of climate change within the last dozen years. These studies explored how people locally make sense of global climate change (Barnes/Dove 2015; Hulme 2016; Jasanoff 2010; Krauss/von Storch 2012), mapped the controversies, inactions and opportunities linked to global warming (Hulme 2009; Storch/Krauß 2013), and specifically focused on the aspects of futurity and invisibility (Doyle 2009; Nicholson-Cole 2005). Building on such new, more culturally grounded understandings of climate change, communication and media studies have established 'climate change communication' as a new distinctive field, taking into account the particularities of climate change for human cognition and sense-making (Moser/Dilling 2007; Neverla 2012; Schneider/Nocke 2014; Sheppard 2012). The *Anthropocene* is a concept which enabled the interment of some of the disciplinary science wars fought around climate change (Crutzen 2006; Steffen et al. 2011):

> The term Anthropocene suggests: (i) that the Earth is now moving out of its current geological epoch, called the Holocene and (ii) that human activity is largely responsible for this exit from the Holocene, that is, that humankind has become a global geological force in its own right. Since its introduction, the term Anthropocene has become widely accepted in the global change research community, and is now occasionally mentioned in articles in popular media on climate change or other global environmental issues. (Steffen et al. 2011: 843)

The Anthropocene concept, since its evocation by climate scientist Paul Crutzen and others, has been key to establishing a new thinking about climate change, enabling more holistic views of human-environment relationships and climate change in particular.[18] Recent achievements incorporating such broader perspectives are reflected in the idea of *planetary boundaries*:

> Since the Industrial Revolution, a new era has arisen, the Anthropocene, in which human actions have become the main driver of global environmental change. This could see human activities push the Earth system outside the stable environmental state of the Holocene, with consequences that are detrimental or even catastrophic for large parts of the world. [...] To meet the challenge of maintaining the Holocene state, we propose a framework based on 'planetary boundaries.' These boundaries define the safe operating space for humanity with respect to the Earth system and are associated with the planet's biophysical subsystems or processes. (Rockström et al. 2009: 472)

The planetary boundaries concept has been developed by a group of interdisciplinary scientists, which include several leading figures of the PIK, such as Hans Joachim Schellnhuber (former director) and Johan Rockström (acting director). The aim of the concept is to 'close the loop' between human and environmental dynamics and integrate all relevant relationships into the methodological framework of Earth System Science (Donges et al. 2017; Schellnhuber/Wenzel 1998). The way to handle such cybernetic integration is to formalize, represent and operationalize them in simulation models. Such dynamic simulations can then be used to 'run into the future,'[19] enabling statements about future risks and to propose modes of steering into more sustainable ways of living.

---

18 While the Anthropocene concept has been able to assemble researchers from the natural sciences, the social sciences and the humanities, it has also fueled new controversies. Among other things, it has been argued that the Anthropocene as a cybernetic concept helped to break the taboo of climate engineering as a solution to the climate crisis (Asayama et al. 2019).

19 A figure of speech in the community of simulation modelers. See, for example, https://bonnsustainabilityportal.de/de/2019/06/fona-erwarmung-der-arktis-fuhrt-zu-wetterextremen-in-unseren-breiten-awi-forscher-entwickeln-

Climate modeling and simulation

Simulation modeling has long been established and represented as the fundamental organizing principle for the global epistemic community that surrounds the climate change issue (Edwards 2001: 34; Sundberg 2007: 473). Accordingly, computer models, simulations and the scientific practices around them have been a recurrent theme in academic fields, such as science studies, STS, and the philosophy, sociology and anthropology of science. A rich literature in STS and the philosophy of science is available addressing epistemic and representational issues regarding climate models and simulations (Gramelsberger 2008a; Pias 2008; Winsberg 2010) and to computer models in general (Sismondo 1999). Along this thread, Gabriele Gramelsberger characterizes the relationship between fish in the ocean and simulated fish in a simulated ocean concisely as follows:

> But these fishes are to be enjoyed with caution, they cannot be angled. You would not even see them if you were diving in the simulated ocean, because neither the ocean nor the shrimp exist in the form we know. Rather, they are semiotic objects, all of which are mathematical in nature and subject to an unimaginable logic and purely functional point of view. (Gramelsberger 2008b: 84, translated by the author)

Gramelsberger and others have aptly characterized the way scientists have translated the world into mathematical models and simulations. Equally, a number of authors have characterized the way computer modelers understand and deal with these complex relationships between models and realities; in climate research (Lahsen 2005; Sundberg 2008) and beyond (Leonardi 2012; Turkle 2009). Many of these accounts have observed a "lure of the virtual" (Bailey et al. 2012) in simulation work – the situation when actors become immersed in their

---

klimamodell-das-den-schwachelnden-jetstream-erklart/, retrieved on May 3, 2019.

virtual worlds and have trouble to distance themselves from their mathematical representations of reality. Myanna Lahsen, for example, contends that "Critical distance is [...] difficult to maintain when scientists spend the vast majority of their time producing and studying simulations, rather than less mediated empirical representations" (Lahsen 2005: 908), and a modeler "explained the difficulty of distinguishing a model from nature [...]" (ibid.: 909).

While I have personally witnessed several such situations of the "lure of the virtual," the relationship between modelers and their simulations at Potsdam Institute differs to some extent from Gramelsberger's and Lahsen's observations. As a matter of fact, I found the scientists surprisingly conscious and reflective about the limits of models in representing phenomena. This difference between my observations and those described in existing literature may be explained by different circumstances: The 2000s was the time when STS scholars and philosophers of science became particularly interested in climate simulation. Climate science had been characterized as the new lead discipline for the dawning age of simulation-driven science (Gramelsberger 2008a: 105). Climate modeling became a focus of newly established research programs in the social sciences and humanities: In Germany, for example, the scientific network *Atmosphere & Algorithms*[20] and the Institute for Advanced Study in *Media Cultures of Computer Simulation.*[21] One reason for this interest in the 'social' and the 'cultural' in climate simulation was that these models became literally overwhelming. The spatial resolutions of the models improved significantly, the phenomena represented became ever more diverse and universal, and its political anchoring and backing increasingly powerful. The 4th assessment report

---

20 https://www.geisteswissenschaften.fu-berlin.de/en/v/atmosphere-algorithms/index.html, retrieved on April 3, 2019.

21 https://www.leuphana.de/en/dfg-programme/mecs/about-mecs.html, retrieved on April 3, 2019.

of the IPCC published in 2007 introduced the visual representations of climate simulations into mainstream media (Mahony/Hulme 2014; Schneider 2012; Schneider/Nocke 2014; Walsh 2014). This body of literature mostly investigated the making of global models, which then became a matter of collective work distributed between countless scientists and infrastructures. These models were all-embracing. Many scientific careers were built and maintained accompanying the life of these models. Scientists spent many years with one global climate model, or rather the representation of a particular aspect within one model. This overlap of a model with one's daily practice and reputation produced risks of immersion into one's own virtual creation.

In fact, this differs considerably from the scientific practice witnessed during my fieldwork at the Potsdam Institute. On the one hand, its scientists work mostly on different models and projects simultaneously. Impact models are relatively small and simple compared to global climate models. As a result, the possibilities and dangers of immersion seem relatively small. On the other hand, the scientists at PIK spend a considerable time with representational work, rather than just writing computer code. Given the increasing social, economic and political relevance attributed to climate research, the scientists have become professional mediators of their work. They have become open scientists, in the sense of a professionalization of representational practice. Regarding the analysis of work at the boundaries of traditional science, it is not useful to treat discretized computer worlds as isolated from the everyday scientific practice of their producers, users and stakeholders. As Cornelius Schubert has argued,

> social performativity might actually have a larger impact on the creation of societal futures than epistemic performativity itself, because it is the social processes of legitimation and justification in which – in a pragmatist sense – the predictions are *'made true.'* (Schubert 2015: 5)

As will be shown throughout this study, it seems convenient to investigate the social performativity of climate impact predictions as a mutual configuration of analog and digital practices within an entanglement of sociotechnical infrastructures. At times, everything seemed permeable and fluid in this world – the facts and artifacts, the communities and social worlds, and the technologies and infrastructures. However, this perceived fluidity may be more of a consequence of our human sensorium and our investigative practices as researchers than of the phenomena themselves. Perhaps, we just need different eyes to look at things.

Simulating climate futures in Germany

I had the status of a visiting scientist at research domain (RD) IV addressing '*Transdisciplinary Concepts and Methods*' in my one-year fieldwork at the PIK. At the time, the domain was an assemblage for researchers and research subjects that did not fit into the other three research domains. This included fundamental research on nonlinear physics, cross-sector activities, such as visualization methods and tools, sociological research and transfer projects developed together with external partners. Accordingly, the annual meeting of the research domain was referred to as 'chaos days,'[22] which has been an apt characterization for the event. The research domain and its annual meeting have now been rebranded as 'complexity science' and 'complexity days.' It seems that researchers have found a way to achieve a disciplined lack of clarity, as John Law put it: "Clarity doesn't help. Disciplined lack of clarity, this may be what we need" (Law 2007: 2). As a matter of fact, this might be an apt description for all endeavors of interdisciplinary and transdisciplinary research, including Earth System Analysis (ESA) and media ethnography. Against this background, the present chapter is a dialogue between climate impact research (research object) and ethnography (method, theory) on tactics aiming at such a disciplined

22 'Chaos days' is the name of an annual punk gathering in Germany.

lack of clarity. It discusses ways of gaining access to, producing data about, and being able to analyze and describe a field of interest. In the case of the PIK, this field of interest is the future with climate change. Within my own research, it is the practice of describing such futures. In both cases, it should be highlighted that the description of the field is a matter of co-construction. In the case of research at the PIK, this translates to a cautious consideration of the epistemic status of models and simulations, which are not to be confused with true representations of a (future) reality. As Sergio Sismondo has rightly put it:

> Whereas theories, like local claims, can be true or false, models and simulations are typically seen in more pragmatic terms, being more or less useful, rather than more or less true. (1999: 247)

This reservation regarding the status of truth in computer simulations is well established at the PIK. The cautious characterization of predictive statements can be illustrated by a seminal scientific paper co-authored by the PIK researchers which describes future scenarios as follows:

> Socio-economic and emission scenarios are used in climate research to provide plausible descriptions of how the future may evolve with respect to a range of variables including socio-economic change, technological change, energy and land use, and emissions of greenhouse gases and air pollutants. (van Vuuren et al. 2011: 6)

Quantitative scenarios have been the primary vehicle for climate research since the 1990s to describe such plausible descriptions of futures.

Regarding the case of contemporary ethnography, the anthropologists Akhil Gupta and James Ferguson have highlighted that its purpose is not to describe the characteristics of a *bounded field* as a representation of a certain reality but rather to investigate *shifting locations* by means of observation, reflexivity and intervention (1997: 138). These shifting locations equally include those of the elements identified within the fieldwork and the position of the researcher (Haraway 1988; Rose 1997). Accordingly, the aim here is not to draw a complete image of the predictive practices at the PIK but to discuss a collage of selected impressions and highlight possible methodological access points for media ethnography. In so doing, the characterization aims at grasping crucial aspects of futurework (Fine 2007: 102) at the PIK: Its heterogeneity and constant alteration, paralleled with a quest for stabilization and harmonious representation.

An idioculture of futurework

Based at RD IV, I had an opportunity to move freely between different research groups and to be fairly independent of the more disciplined agendas and working routines at the rest of the institute. When I entered the PIK, I was particularly interested in figuring out how researchers at this specific location are able to produce scenarios about the future with climate change. While such knowledge about the future is not the only matter of concern for the institute, it is certainly its specialty and the origin of its reputation. The PIK's idioculture is one of futurework. The sociologist Gary Alan Fine has coined the term "idioculture" for the way a setting of cultural elements defines how members of a community interact with each other at work or generally in life:

> Idioculture consists of a system of knowledge, beliefs, behaviors, and customs shared by members of an interacting group to which members can refer and employ as the basis of further interaction. (1979: 734)

Fine has conducted a number of fascinating studies of idiocultures, including those in little league baseball teams (ibid.), fantasy role-playing communities (Fine 2002), meteorologists (Fine 2007) and restaurant workers (Fine 2008). The study on Chicago-based meteorologists, *Authors of the Storm: Meteorologists and the Culture of Prediction* (Fine 2007), is especially relevant in the context of this study on climate modelers. As Fine argues, different idiocultures of work are occupied with different temporalities. Some jobs are dealing with the temporalities of the past (e.g. librarians, historians, archeologists), many have a strong connection to the present (care workers, news journalists) and some have a deeper relationship with the future.

> [...] a few are given the assignment of looking forward, such as physicians, financial planners, fortune-tellers, pollsters, and, here, meteorologists. They engage in futurework. (ibid.: 102)

This does not mean that futurework can discard the past and present, only that these temporalities are interpreted towards a definition of the future:

> These boundaries are not hard and fast, but a matter of emphasis. A police detective might be charged with determining the location at which an arrest can be made or may be asked whether a criminal is likely to strike again; the car salesman, about a repair history of a vehicle or the likelihood of a car needing repairs in the future; the internist may be asked about the meanings of past symptoms, current medical interventions, as well as the patient's prognosis. (ibid.: 102)

Fine's analysis inspired an international network of interdisciplinary researchers[23] to think about distinct cultures of prediction in climate modeling. The anthology *Cultures of Prediction in Atmospheric and Climate Science,* the principal outcome of the research network, aims at offering a "broadened framework of cultures of prediction to describe and better understand postwar predictive efforts based on computer simulation" (Heymann et al. 2017: 19) The articles cover topics such

23 Atmosphere & algorithms, a scientific network funded by the DFG German Research Community (2010–2012). https://www.geisteswissenschaften.fu-berlin.de/en/v/atmosphere-algorithms/index.html, retrieved on April 19, 2019.

as the transformation of climate models from heuristic to political instruments (Heymann/Hundebøl 2017), the downscaling of global climate models to the local level (Mahony 2017) and the visual semantics of the future in climate change imagery (Schneider 2017). Building on these case studies, the editors of the anthology highlight a number of key characteristics that drive predictive practice in climate research more generally. These include the important social role of the predictions, the character and significance of computational practices, the domestication of uncertainty, the degree of institutionalization and professionalization of predictive expertise, and the cultural impact of predictive practices and claims (Heymann et al. 2017: 20). However, the case studies of the anthology also show how idiocultures of prediction differ from place to place and from situation to situation. In the following, I will discuss some aspects that differentiate the predictive culture and practice at the PIK from those of other places and actors in the field of climate research.

Towards Earth System Analysis

In the case of the Potsdam Institute, the specialization in futurework has been an essential component of its institutional design. We can identify some of the characteristics of this 'futurework by design' by going through a conceptual anthology *Earth System Analysis: Integrating Science for Sustainability* published by PIK director Hans Joachim Schellnhuber and others in 1998. In the introduction to the anthology, the editors Hans Joachim Schellnhuber and Volker Wenzel outline the ingredients of ESA, as a "science *in statu nascendi*," having

> 1. a genuine subject, namely the total Earth in the sense of a fragile and "gullible" dynamic system,
>
> 2. a genuine methodology, namely transdisciplinary systems analysis based on, i.a., planetary monitoring, global modelling and simulation,

> 3. a genuine purpose, namely the satisfactory (or at least tolerable) coevolution of the ecosphere and the anthroposphere (vulgo: Sustainable Development) in the times of Global Change and beyond. (Schellnhuber/Wenzel 1998: vii)

The editors also reflected on the process of scientific innovation in a broader sense, including the obligatory reference to Thomas Kuhn and his paradigm shifts:

> Employing T. Kuhn's all too popular epistemological theory, we have to search for a generating paradigm shift as triggered by some "experimentis crucis", a major historical event, certain revolutionary technological developments, or the like. (ibid.: viii)

The rhetoric artfully plays with temporalities, considering the paradigm shift towards ESA as a realized event in the past, thereby confirming its existence and relevance. For the authors, the birth of ESA is triggered by the "co-operation" of three crucial factors: Firstly, "the race for the Moon created the opportunity to observe planet Earth from space with sophisticated equipment" (ibid.), which made ESA "conceivable." Secondly, ESA became "feasible" through

> the advent of electronic super-computers established the technological platform for sufficiently fast and comprehensive global simulation modelling based on adequate management of the plethora of now available monitoring data. (ibid.)

Thirdly, the discovery of the ozone hole confronted the international community "with the evidence that humanity can and, in fact, is about to transform the character of the global environment," (ibid.) thereby, making ESA "mandatory." The main part of Schellnhubers conceptualization focuses on the element of technical 'feasibility,' emphasizing the role of computers and simulation experiments in ESA:

> the advent of sophisticated parallel computer hard- and software [...] in combination with recent progress made in scientific modelling of complex systems might allow the establishment of *virtual impact laboratories*. Renewable artificial Earth systems could be exposed there to various simulated crash scenarios in order to study the potential consequences. As a matter of fact, these cyberspace experiments should be the most powerful tool for generating entire ensembles of assessments within a reasonable stretch of time. (Schellnhuber 1998: 8, emphasis in original)

Schellnhuber acknowledges the origin of computer simulation in military contexts and advocates for further repurposing of these technologies for the social good: "Why shouldn't we make full use of the knowledge and methods involved for less destructive purposes like the preservation of our environment?" (ibid.: 133) The text particularly demonstrates Schellnhuber's focus on what-if thought experiments and engagements with the future. Accordingly, he describes the challenge of climatic change with an analogy of a predicted meteor collision and its impacts on Earth:

> Imagine ... that astronomers were warning us of a huge asteroid heading towards our planet. The collision was supposed to occur in some twenty years from now, but neither the date of the impact nor its site could be predicted with satisfactory precision at this point in time. From the already available approximate knowledge of the celestial maverick's mass and orbital parameters it could be inferred, however, that the collision energy would correspond to an explosion of at least 10 gigatons of TNT. (ibid.: 5)

Schellnhuber elaborates on questions regarding the probability of such an event and then goes further, asking – what would mankind do? According to him, humankind would come together and ask the scientific community to work out a comprehensive impact analysis assessing aspects such as collision probability distribution, identification of consequences, options for protection, adaptation or rehabilitation and options for mitigating or even preventing the collision (ibid.: 6) Adapted to the issue of climate change, this is a fairly detailed description of the challenges scientists at the PIK have been dealing with since the establishment of the institute in 1992 – the translation of human-environment relationships into discrete classes and mechanisms, the calculation of probabilities of future impacts of climate change, and the proposition of ways to deal with these consequences by mitigating GHGs and adapting to the unavoidable.

Earth System Analysis highlights the quality of control through all-compassing information well beyond the scope of traditional climate science. Schellnhuber was explicitly influenced by the idea of cybernetics and suggested an operationalization of sustainability management as *geocybernetics,* described in an article as “the art of adequately controlling the complex dynamic Earth system under uncertainties of all kinds” (Schellnhuber/Kropp 1998: 411). The issues at stake for the PIK director require a radical rethinking of traditional methods used in the sciences, challenging Karl Poppers principles of hypothesis testing and falsification: “[...] We are not willing or allowed to sacrifice the integrity of the one and only planetary specimen we have got for the sake of scientific progress” (Schellnhuber 1998: 131). Against this background, he proposes simulation modeling as a tool for virtual experimentation, a generation of future projections and a proposition of management options:

> There is one way out of this dilemma, however, namely virtual falsification or verification of Global Change hypotheses with the help of artificial copies of the Earth System or of crucial parts of the latter. [...] a runaway greenhouse event, e.g., in virtual computer reality may cost us one CPU year on the most advanced CRAY machine, but not our lives! In such a case, we simply restart the digital game and try to employ a more careful strategy. In this way we may eventually be able to explore the plume of potential coevolutionary futures as generated from the present state of the Earth System by the management options contained in $\mathfrak{N}$. (ibid.: 133)

Keeping the origins in cybernetic thinking in mind, it becomes clear that work carried out at PIK has always been more than traditional science. Andrew Pickering has described the relationship between traditional science and cybernetics very adequately as follows:

> While classical science has thus been an epistemological project aimed explicitly at knowledge production, cybernetics is an ontological project, aimed variously at displaying, grasping, controlling, exploiting and exploring the liveliness of the world. (2002: 430f)

The geocybernetic concept for ESA is also reflected in the original logo of the Potsdam Institute, depicting a tree (Earth system), a troubled sky (climate change), and a human hand (human agency), (en)framed in a purple triangle (ESA) (see Fig. 11).

Figure 12: Old (left) and new (right) logo of the Potsdam Institute. Source: PIK

As we will see later on, computer modeling and simulation may be characterized as the framework (cf. Heidegger 1954) stabilizing the Pandora's Box of transdisciplinary sustainability research. The emblem was later replaced with a logo depicting the PIK headquarters and Spieker architecture formerly hosting the astrophysicists of AOP. One may read this as a sign of institutional maturation and subordination to the traditions of the Science Park Albert Einstein.

Disciplining transdisciplinarity through technology

One of the features of work at PIK is its truly transdisciplinary nature. As Andrew Barry, Geogina Born and Gisa Weszkalnys have shown,

> ideas of interdisciplinarity and transdisciplinarity imply a variety of boundary transgressions, in which the disciplinary and disciplining rules, trainings and subjectivities given by existing knowledge corpuses are put aside or superseded. (2008: 21)

Despite the instability resulting from inter- and transdisciplinary settings of scientific work, such a practice has been hailed as a solution to a variety of perceived contemporary problems for many years. This includes the relationships between science and society, the development of accountability and the need to foster innovation in the knowledge economy (ibid.: 21). Accordingly, interdisciplinary work may be focused on very diverse issues and manifest itself in a variety of ways. Interdisciplinary researchers may gather around so-called grand challenges of humanity (sustainability studies, climate research), around new sociotechnical configurations,[24] or around methodological devices (ethnography, artistic research). Referring to Nigel Thrifts *Re-inventing Invention* (2006), they invoke the example of ethnography in the IT industry, which

> offers a set of techniques through which businesses are expected to be able to transform their knowledge of and engagement with those micro-spaces of

24 E.g. computer-supported cooperative work.

> social life, replete with social and cultural difference, to which they previously did not have access […]. (Barry et al. 2008: 32)

These techniques may then create new forms of technical objects that are recognized as, at once, socially and culturally embedded (ibid.: 25). Regarding disciplinary composition, a common constellation is a collaboration between natural sciences or engineering, on the one hand, and the social sciences, humanities or arts, on the other (ibid.: 228). The power, relationships and roles of the different disciplines involved is often a matter of fierce debate in the conceptualization and practice within interdisciplinary settings. Barry and colleagues characterize these different modes of interdisciplinary engagement as follows: (1) The integrative-synthesis mode, where two or more disciplines essentially merge together. Examples for this constellation are biochemistry and astrophysics. (2) The subordination-service mode. Here, disciplines are organized according to a clear hierarchical division of labour, with one or several disciplines providing services or complementing the work of others. An example of such collaborations are ELSA-arrangements, where social scientists are invited into projects within the natural sciences to comment and critique on ethical, legal and social aspects (i.e. 'ELSA') (see Hullmann 2008). Similarly, settings in the Digital Humanities often employ scientific programmers (computer scientists), who provide technical services to the project-leading humanities researchers. In both cases, these disciplinary roles may be seen as problematic (see for example Balmer et al. 2016 for a problematization of ELSA/ELSI). (3) The agonistic–antagonistic mode, where "interdisciplinarity springs from a self-conscious dialogue with, criticism of or opposition to the intellectual, ethical or political limits of established disciplines or the status of academic research in general. (ibid.: 29)

Depending on the level of dissolution of disciplinary boundaries and the integration towards the new subject, the cooperative settings may be framed as 'interdisciplinary' (agonistic–antagonistic, subordination-service) or 'transdisciplinary' (integrative–synthesis). In other cases, collaboration may go well beyond the boundaries of scientific fields, therefore, evoking characterizations of post- and a-disciplinarity (Krishnan 2009; Sayer 2000).

As we have seen concerning the example of the AOP, different aspects are conducive for the establishment of innovative scientific work beyond established disciplines. Obviously, the creation of a formal institution enables the mobilization and reception of financial funds, mostly emanating from fiscal resources in our case. The foundation of AOP as a formal institution in 1874 was followed by an allocation of financial funds to be spent for construction materials, wages and technology on Potsdam Telegrafenberg. The institutionalization was also necessary as a means of national recognition and admission into the international networks of astronomy. However, institutionalization alone does not explain how astronomers, physicists, mathematicians, and technologists were able to collaborate and steadily develop a routine later becoming recognizable as astrophysical practice. Rather, it was precisely the decision to establish a new, particular and independent place *to do astrophysics* that enabled the development of this new practice. As highlighted before, it would have been logical to base the new institution and initiate astrophysical work in the center, Berlin. However, drawing together people, technologies and infrastructures on this formerly blank sheet of Potsdam Telegrafenberg facilitated a new perspective on the night sky, the Earth, science and the use of technological instrumentation. By shifting locations (Gupta/Ferguson 1997) from Berlin to Potsdam, one might forget existing mental maps and unlearn established routines that prevent such new views. On the other hand, unlearning may not only have taken place but also taken time. It was helpful that the AOP scientists and technical personnel were not only closely working together but also living there together at

Telegrafenberg. Spending time together enabled them to tinker with new kinds of observatory technology and methods. These experiments then finally stabilized around the method of spectral analysis, which has guided astrophysical research practice until today.

Similarly, transdisciplinary work at the PIK is primarily structured around computer modeling and the digital infrastructures of computational science. This aspect of the scientific method as a stabilizing element for work at the PIK is echoed by the mission statement on its institutional website:

> PIK addresses crucial scientific questions in the fields of global change, climate impacts and sustainable development.
> Researchers from the natural and social sciences work together to generate interdisciplinary insights and to provide society with sound information for decision making.
> The main methodologies are systems and scenarios analysis, modelling, computer simulation, and data integration.[25]

While paragraph one opens up a Pandora's Box of heterogeneous challenges and fields and paragraph two gathers the skills of heterogeneous actors to address these challenges, paragraph three stabilizes this work by means of a limited methodological toolbox: "[S]ystems and scenarios analysis, modelling, computer simulation, and data integration."

Programming the future

The expertise of the PIK researchers in computer modeling, simulation and digital analysis is the most notable common denominator at the PIK. Digital calculation does not only structure the scientific working practices at the institute, it is a genuine element of its idioculture. This omnipresence of computer code at the PIK can be illustrated by a note

25 https://www.pik-potsdam.de/institute/mission/mission?set_language=en, retrieved on April 2, 2019.

hanging in the shared kitchen of the House of the Woods. The note urges colleagues to keep the kitchen clean and to wash dishes by hand if the dishwasher is out of service. The note is written as computer code in the programming language Python, which proposes a way to 'run' the kitchen (see Fig. 13).

Very broadly, scientific practice at the PIK has been interwoven with the development and mainstreaming of computational sciences in the last few decades. Gabriele Gramelsberger traced the role of computers for scientific practice historically in her book *Computerexperimente.*

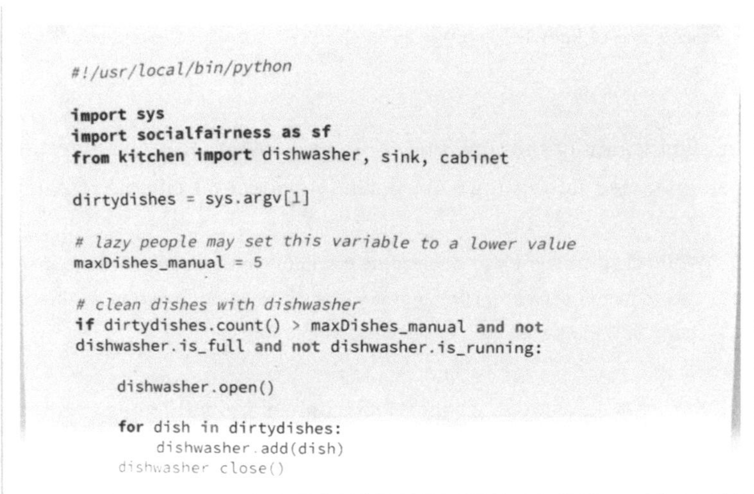

```
#!/usr/local/bin/python

import sys
import socialfairness as sf
from kitchen import dishwasher, sink, cabinet

dirtydishes = sys.argv[1]

# lazy people may set this variable to a lower value
maxDishes_manual = 5

# clean dishes with dishwasher
if dirtydishes.count() > maxDishes_manual and not
dishwasher.is_full and not dishwasher.is_running:

    dishwasher.open()

    for dish in dirtydishes:
        dishwasher.add(dish)
    dishwasher.close()
```

Figure 13: Dishwasher Python note in kitchen.
Source: unkown, picture taken by the author at PIK kitchen

She describes how the appearance of electronic computers and their worldwide distribution changed the practice of and environment for scientific knowledge production massively: "No discipline and no scientific method remain unaffected by the use of computers"

(Gramelsberger 2008a: 85, translated by the author). For her, the advent of computers in science marks "the second half of the scientific revolution in modern times" (ibid.: 85). Computational departments have emerged in almost all scientific disciplines in the last fifty years. In addition to theory, expertise, observation and measurement, simulation was added as a new epistemological instrument in the 1950s, and has been used more intensively since the 1970s.

> The newly established computational departments follow their own research logic, which is characterized by the numerical analysis of complex systems, application- and problem-oriented research questions, a high degree of interdisciplinary cooperation and international networking as well as the dependence of knowledge progress on the performance of computers. (ibid.: 96)

The significance of the computer as an instrument for research, experimentation and forecasting is particularly evident for climate research:

> While the astronomy was the leading discipline in the transition from medieval to modern science, climate research is the leading discipline in the currently developing simulation sciences. (ibid.: 105)

Daily practice in computational science only marginally intersects with the 'BC' (before computers) practices of the originating disciplines. This discrepancy between the daily work at the PIK and the scientific traditions at university was a salient issue in my interviews with scientists at the PIK. As researcher Jeremias Scholz explained to me, he originally studied physics but says that he learned most of his professional skills during his PhD studies at the PIK. When I asked him what he referred to, he mentioned computer programming in particular. He had only taken one single course in C++ at university. He then did his PhD at the Potsdam Institute, acquiring more comprehensive skills in C++, C and FORTRAN. As Gabriele Gramelsberger argues, these computer languages and their codification of mathematical formula can be described as a *lingua franca*, without which the interdisciplinary collaboration in computational science would be inconceivable (Gramelsberger 2008a: 144). As a matter of fact, many mathematical equations, models and program parts can be found in different disciplines. The Navier-Stokes equations of fluid dynamics, for example, are equally used in atmospheric and ocean models, as well as in medical and technical fluid simulations (ibid.). Scientific programming uses theory in a modular way fit for experimentation. It comes in the form of a construction kit composed of programmed theoretical building blocks (ibid.).

According to Jeremias, many modelers work with the Network Common Data Form[26] (NetCDF) infrastructure, a set of software libraries and self-describing, machine-independent data formats that support the creation, access and sharing of array-oriented scientific data. Younger researchers especially also use the programming language Python, which essentially builds on open-source codes and communities (see Chapter IV).

> Once, I had a database in Excel with information on city names and I wanted to plot that on a map. So, I looked for a library for Python to do this. And I found a free library to do this on the web. The place to look for these things is Github. There are some researchers who often publish on Github and there are automatic programs that create a documentation for your code to publish it on the Github website. (Discussion with Jeremias Scholz during lunch at Telegrafenberg Cafeteria)

I asked Jeremias if the PIK modelers also use cloud services. He told me that many modelers work with data on the Earth System Grid,[27]

26 On NetCDF: http://www.unidata.ucar.edu/software/netcdf/, retrieved on July 14, 2019.

27 On Earth System Grid: https://www.earthsystemgrid.org/home.html, retrieved on July 14, 2019.

which is stored by the Deutsches Klimarechenzentrum[28] in Hamburg. The Earth System Grid is an international data distribution portal that dispenses information for the IPCC Assessment Reports, the most important international publication format for policy-oriented climate research. Otherwise, the PIK has a very strict data policy and does not allow scientists to use commercial cloud services (e.g. Amazon Web Services, Google Cloud, IBM Cloud) for computing and storage. As a matter of fact, my interviews with US researchers also suggest that this is a German (and European) cultural specificity. United States scientists are generally more open to using services from private companies, be it software tools (e.g. Esri's ArcGIS) or cloud services (e.g. AWS). Permeability between scientific and commercial worlds is also facilitated by cultural and geographic proximity, given all of the relevant services are provided by US companies. We can give the example of the 2016 annual meeting of the American Association of Geographers, where the author and colleagues from Locating Media participated with presentations. The gigantic conference featured the company Esri as prominent main sponsor, a situation unimaginable in Europe.

Working in technology

The reliance on digital infrastructure at the PIK is reflected by the interior design of the offices at the House in the Woods. My furniture included a desk, a Dell desktop computer, a telephone, a plastic daisy flower, a bookshelf, and a number of empty book spines made of

28 On Deutsches Klimarechenzentrum: https://www.dkrz.de/, retrieved on July 14, 2019.

cardboard. Tim Neitzel commented sarcastically on the impertinence of installing a humanities scholar in an office with fake books.

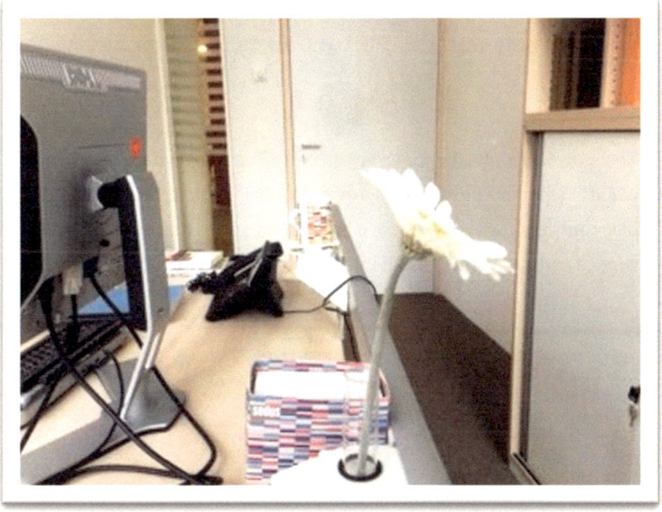

Figure 14: Model office in the Earth wing of the House in the Woods. Source: Own photo

Books seem to vanish from the office spaces at the PIK. Some senior scientists still host entire personal libraries in their rooms, but the equipment and furniture of younger scientists is mostly limited to a laptop and one or two monitors. Computers are omnipresent at the PIK. Knowing how to use the digital infrastructure is key to become part of the PIK ecosystem, as the following episode illustrates. Just after entering my new office, I started the desktop computer whose system was under my table. After an hour, it became clear to me that I had dropped a brick. The supposed PC had not been a PC. Instead, it was a remote visualization server, which had been stored in the formerly empty office. Someone later put warning signs on the servers as I had not been the first to make the mistake (Fig. 15).

Star and Ruhleder reminded me that infrastructure is always learned as part of membership to a community of practice: "Strangers and outsiders encounter infrastructure as a target object to be learned about. New participants acquire a naturalized familiarity with its objects as they become members" (1996: 113). And as I could learn during the first two hours at the PIK, infrastructure becomes visible upon breakdown:

> The normally invisible quality of working infrastructure becomes visible when it breaks; the server is down, the bridge washes out, there is a power blackout. Even when there are back-up mechanisms or procedures, their existence further highlights the now-visible infrastructure. (ibid.)

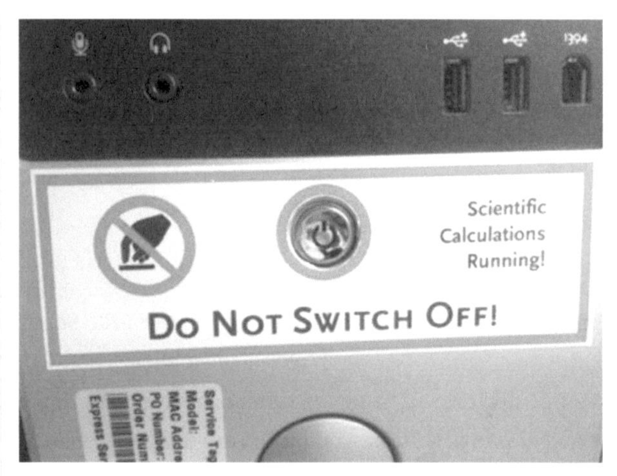

Figure 15: Do not switch off! Scientific Calculations Running! Source: Own photo

The beating heart of the digital infrastructure at the House of the Woods is the supercomputer living in the basement of the building. The 'cluster,' as such high-level performance computers are referred to by actors in the community, had been installed as an integral part of the architecture finalized in late 2015. Such supercomputers are prestige objects that are gladly financed by governments, in this case by the

German Federal Government, the Federal State Government of Brandenburg and the European Union. Tragically, they have a rather short life and are typically outdated after five years. At the time of acquisition (2015), the computer featured ranked 353 of the world list of the fastest computers.

Characterizing the supercomputer as the 'beating heart' is more than a metaphor. The House in the Woods is a materialization of the 'organic architecture' imaginary, promised (but not realized) by the Einstein Tower. The shell of the building is made of wood paneling, which allows for a perfect embedding into the forested environment of Telegrafenberg. Looking outside the window of the office, one can observe deer and rabbits on a daily basis. The form of the House in the Woods is also 'organic,' representing a three-leafed clover, which can be spotted from space. Similar to the Einstein Tower, the outer and inner design of the House in the Woods avoids right angles and has the feel of a fluid sculpture rather than solid, stable architecture. Entering the building, one stands in an atrium flooded in light. Offices and meeting spaces are distributed on three floors and the three wings (cloverleaves), which are designated as the Earth, the sky and the sun.

Figure 16: View from my office into the woods (left) and inner atrium (right). Source: Own photos

The meeting rooms bear the names of geographic continents, highlighting the global dimension of climate impact research. Given the omnipresence of computers in the work of the PIK scientists, it may be surprising that digital infrastructure is virtually invisible to the human eye. The only visible trace of digital infrastructure for the outsider is the signpost (Fig. 17) next to the elevator, featuring the supercomputer in the basement. One seems to gain direct access to the cluster by pressing the button for level -1, which is, of course, an illusion. The server room is well locked and can only opened by a few dedicated employees who have been on board the PIK since its beginning.

In fact, we should consider the 'black boxing' of technology and infrastructure as a conscious design choice. We will see later that the PIK's work is not just about calculation, but a lot about accountability and representation. Technology is understood as the primary working tool enabling statements about climate change and the future. By contrast, the obligatory passage point (Callon 1984) to mediate the insight of these technological experiments to outsiders is the PIK scientist, possibly without the consideration of help from nonhuman elements. It is telling that the supercomputer (calculation) and the conference room (accountability) share the same floor in the House in the Woods.

Figure 17: Floor plan of the House in the Woods. Source: Own photo

The supercomputer is connected via sophisticated network structures and can technically be accessed from all the offices at the House in the Woods. In this sense, the House in the Woods resembles a cybernetic organism, consisting of material (machines, cables), human (scientists) and symbolic (distributed network software) elements. In this context, we can again refer to Thomas Gieryn's article "What buildings do," in which he proposes analyzing buildings as "walk-through machines":

> A different sense of buildings comes from seeing them as 'walk-through' machines. Buildings are technological artifacts, made material objects, and humanly constructed physical things. To see them this way brings buildings within the compass of a promising theoretical orientation developed initially for the study of machines. (2002: 41)

Having the example of the PIK building in mind, this methodological approach gains another quality and significance. Buildings in the age of smart architecture actually *are* machines, in the common sense of the word. Equally so, computers become walk-through architectures again, as in the early days of the technology. The machinist nature of the architecture is encompassing but can be described specifically by the following conjunction of elements. As a matter of fact, the supercomputer is not only a conditional element for the scientific experiments carried out at the PIK but also provides all the heating energy in winter. The entanglement of different energetic circuits (information, heating, cooling) and media (electric energy, water) creates opportunities but also new risks. As Klaus König, the PIK's head of IT systems highlights:

> There is one disadvantage of this direct cooling. If there is a breakdown in the water circuit, I also have to switch off the cluster to be able to repair it. (Interview with König)

The only incident when Klaus König had to shut down the supercomputer since its installment was triggered by such a problem in the cooling system. One of the filters in the water circuit had been clogged by bacteria in the water, thus, needing replacement.

Figure 18: Fluid elements running through the system. Vertical tubes for water, black cables for electricity and blue for information. Source: Own photo

As a consequence, all scientific simulation had to be stopped and the CPUs turned off until the congestion was eliminated. Apart from this onetime exception, the supercomputer has been a valuable and essential companion for the scientists. Nevertheless, the episode shows that the technical integration of several vital functions into one system comes with a considerable increase of complexity, risks of infrastructural breakdown, the need for new control mechanisms and irreplaceable human expertise. One has to gain considerable situated knowledge to understand the behavior of these multiple fluidities running through the computer (see Fig. 18).

Calculation and accountability

It seems obvious to characterize the Potsdam Institute as a center of digital calculation. Bruno Latour has used this term in his book *Science in Action* for sites,

> where inscriptions are combined and make possible a type of calculation. It can be a laboratory, a statistical institution, the files of a geographer, a data bank, and so forth. (1999b: 304)

If we use Latour's term here, it must be added that inscriptions entering the PIK are different from the ones discussed by Latour, being specimens, probes, paper maps and tables. Inscriptions reaching the Potsdam Institute take the form more of standardized, digital datasets. This may entail, for example, a numerical time series projecting a spatiotemporal change of temperatures within the 21st century. They are equally an element and a late product of what Paul Edwards refers to as the "vast machine," a globally distributed climate knowledge infrastructure (2010: 432). As a result, the data have already passed through numerous rounds of refinement and standardization within other centers of calculation. Considering its holistic claim to take into account all dimensions of sustainable development, The PIK is not only processing weather- and climate-related data but also socioeconomic indicators from statistical agencies, natural disaster damage data from reinsurance companies, risk assessment data linked to armed conflicts, and so on. In this sense, the PIK is not only a part of the vast machine described by Paul Edwards but a globally distributed climate knowledge infrastructure (ibid.: 432). It is also an element of a macroeconomic, commercial and international security machine. Or rather, this attribution of infrastructures with occupational fields and social worlds may also have to be reconsidered and reviewed in a world marked by dissolving boundaries between occupational fields, multipurposed technologies and infrastructures and a deep permeability of data (see Chapter V). *Inter alia*, this means that scientists working at the PIK have no control over and fairly limited knowledge about the making of the data in the first place. This situation creates challenges of trust, which are countered by a number of practices, including the standardization of model components, simulation procedures, data formats, and extensive obligations for documentation and reproducibility. Regarding global climate data, Edwards contends:

> To make data global scientists developed suites of intermediate computer models that converted heterogeneous, irregularly spaced instrument readings into complete, consistent, gridded global data sets. They also literally created data for areas of the world where no actual observations existed. [...] As time went on, these techniques became so tightly intertwined that they transformed the very meaning of the term data. Today, the processes atmospheric scientists invented are ubiquitous not only in geophysics but throughout the sciences. Virtually everything we now call "global data" is not simply collected; it is checked, filtered, interpreted, and integrated by computer models. (ibid.: 188)

These standards are required for every scientifically relevant element that enters the institute. Just as well, the PIK ensures that everything leaving the institute meets the same requirements, which is operationalized by internal and external audit and review mechanisms, and through informal peer pressure. As Edwards has argued, scientists themselves are constantly engaging in practices of infrastructural inversion: "The climate knowledge infrastructure never disappears from view, because it functions by *infrastructural inversion*: Continual self-interrogation, examining and reexamining its own past" (ibid.: 432). Such activities of infrastructural inversion are undertaken within the many nodes of distributed networks, such as the vast climate machine. However, what distinguishes centers of calculation from other individual nodes is their ability to co-create the procedural rules for the game. Accordingly, anthropologist Richard Rottenburg has argued in *Far-Fetched Facts*, his parable of development aid:

> The translation chain selected here is connected with many other chains in a worldwide network. Not all of the individual nodes in this network have the same meaning or significance, even if locally appropriate translations exist everywhere. Some nodes are able to define the procedural rules of the technical game in such a way that others are forced to follow them. (2009: 87)

The Potsdam Institute clearly represents such a central node, wielding considerable power within its networks. In so doing, it is not only able to co-create the rules of the game but to redefine who is playing it. This parallelism between calculation and accountability can be illustrated by the making of Shared Socioeconomic Pathways (SSPs) at the PIK. The Potsdam Institute is involved in the process of generating globally agreed quantitative scenarios for GHGs and socioeconomic development in the 21st century. As mentioned before, scenarios have long played an important role in simulation-driven climate research. An increasingly broad array of scenarios had been developed over time. On the one hand, this was due to the fact that more and more issues became the object of scientific scrutiny in climate-related research. Moss and colleagues (2010: 748f) introduce a typology of prominent scenarios used in climate-related research, which can be summarized as follows:

Socio-economic scenarios describe the evolution of the society and ecosystems, in the absence of climate change or climate policies. For example, such scenarios can represent future conditions of economic growth, GDP [Gross Domestic Product] and population size.

Emissions and radiative forcing scenarios. Emission scenarios describe potential future discharges to the atmosphere of substances that affect the Earth's radiation balance, such as greenhouse gases and aerosols. Accordingly, they focus on long-term trends in energy and land-use patterns. Radiative forcing scenarios in contrast express radiative forcing, i.e. potential future changes in energy in the atmosphere due to GHG emissions.[29] According to the nature article, "it is important to differentiate between emissions and RF scenarios, because radiative forcing takes place after a certain time lag following the discharge of GHGs" (ibid.).

Climate scenarios are representations of future climate conditions such as temperature, precipitation and other climatological phenomena.

Impact, vulnerability and adaptation scenarios focus on changes in environmental conditions. While such changes may occur regardless of climate change, the latter often influences them. For example, such scenarios can represent future water availability and quality at basin levels, sea level rise, and characteristics of land cover and use. As such factors often affect the vulnerability of natural and social systems, they can be described and measured by vulnerability studies and scenarios. In the end, such studies can also serve as input for impact scenarios analyzing possible coping mechanisms with the changes ahead. (ibid.: 749)

However, the multiplication of application contexts of scenarios was only one reason for the mushrooming of scenarios in climate research. A second factor was the need to improve existing scenarios by considering new research insights. A third reason to trigger the development of new scenarios was linked to organizational concerns within the community of climate researchers: Namely, the community realized that working on climate issues would only be effective if researchers adhere to a shared set of scenarios. Such harmonization, standardization and disciplining work was understood to increase the consistency and collaboration of climate research and policy. All these factors led to a

29 Radiative forcing (RF) is the measurement of the capacity of a gas or other forcing agents to affect that energy balance, thereby, contributing to climate change.

number of scenario generations that were broadly shared within the climate community: Namely SA90, IS92, SRES (Girod et al. 2009) and the Representative Concentration Pathways (RCPs) and SSPs used currently. The Potsdam Institute is one of the research institutes in this globally distributed effort of developing global community scenarios. In so doing, it co-creates the "scenario matrix architecture" (van Vuuren et al. 2014) that serves as an underlying infrastructure and obligatory passage point (Callon 1984) for all prospective futurework within the community of climate researchers.

Scientists at the PIK are not only contributing to the organization of the climate research community, but they also constantly mediate between the social worlds of climate science and policy. John Schellnhuber was involved in the conceptualization of the IPCC,[30] the United Nations body for assessing the science related to climate change, from early on. This proximity to the world climate council was formative for the PIK and explains the strong influence that the climate modelers on Telegrafenberg exert on international scientific and political processes regarding climate change. The alignment with the IPCC is not only reflected in the thematic focus and personal ties but has been imprinted in its institutional structure. In addition to domain IV (my affiliation), research domains mirror the WGs of the IPCC:

PIK	**IPCC**
Earth System Analysis (RD I)	The Physical Science Basis (WG I)
Climate Impacts and Vulnerabilities (RD II)	Impacts, Adaptation and Vulnerability (WG II)
Sustainable Solutions (RD III)	Mitigation of Climate Change (WG III)

Table 1: Comparison of the PIK's research domains (RDs; left) and the IPCC's working groups (WGs; right). Source: Own table

30 https://www.ipcc.ch/, retrieved on February 3, 2019.

This strategy to focus on the IPCC paid off for the institute in the sense that several PIK researchers have acted as lead authors of the influential assessment reports of the climate council. The IPCC's WG III on *Mitigation of climate change*, for instance, is currently coordinated by PIK researchers.

Investigating the boundaries of openness

The PIK scientists have become equally cautious and professional in the representation and mediation of their scientific work because of the centrality of the PIK in public debates regarding the climate crisis. They became not only experts on calculation but also the accountabilization of their calculations. This professionalization constituted a challenge but also an opportunity for my ethnographic research. I observed and participated in my fieldwork in many of these activities aiming at 'communicating' climate research to outsiders. More than that, I have been heavily dependent on such practices of mediation in order to gain access to information. As ethnographic researchers, we should not take this provision of access for granted but consider our own shifting position towards or within our field. Why have I been given access to an institution or how did I have to position myself in order to gain access to informants, information and infrastructure? As the sociologists Hirschauer and Amman have argued, a certain mimesis of the person, a fit into the milieu, is necessary for the design of the copresence as a 'disturbance' introduced into the field of investigation (1997: 25). In my opinion, this does not only mean that the ethnographic researcher is increasingly able to adapt to his/her 'field.' Rather, the personal biographies and attitudes of researchers towards the field already enable certain access points and prevent others. In my case, I could contribute some insights from past work as a policy consultant for climate policy, I had an interest and some knowhow about map design, and had already cooperated with PIK scientists in a project with the University of Potsdam (Schneider/Nocke 2014). Without this preliminary work, access to the field (in this case, the institution of the PIK) would probably have been impossible.

The identifications, roles and functions ascribed to me in the diverse situations of contact with interlocutors differed considerably. Depending on the context, I was perceived as a visualization designer, knowledge sociologist, media or communication scientist, anthropologist or cultural scientist.

> Search ambivalent identifications, or perceived identifications, immediately locate the ethnographer within the terrain being mapped and reconfigure any kind of methodological discussion that presumes a perspective from above or 'nowhere.' (Marcus 1995: 112)

My investigation during my field research was not limited to observation but increasingly included interventions. I became a mediator of climate change impacts within my role as a visiting researcher at Potsdam Institute; I commented on the graphics and presentations of scientists, created a series of workshops around the 'visualization of climate change' and worked on a publication together with climate scientists. As Marc-Anthony Falzon has pointed out, "[...] ethnographers typically think of data as a gift from their informants, with all the implications of reciprocity that gift exchange implies" (2016: 1).

George Marcus pointed out that the 'mimesis' between the ethnographic researcher and his/her field often entails professional and private spheres:

> In contemporary multi-sited research projects moving between public and private spheres of activity, from official to subaltern contexts, the ethnographer is bound to encounter discourses that overlap with his or her own. (1995: 112)

I became very sensible to the climate debate and how it was conducted within the public sphere during my fieldwork at the PIK. It became apparent to me that it is virtually impossible to separate 'the science' and 'the politics' of climate change. Accordingly, it has to be considered as a political statement and positioning to work at the Potsdam Institute and not at another institute of the Earth Sciences. As a matter of fact, this political demarcation of the PIK and its researchers became especially apparent when I was in contact with researchers of the second geoscientific institute on Telegrafenberg, the GFZ. The relationship between the two institutes (PIK and GFZ) has always been marked by strong competition, sometimes latent resentment and mistrust. On the part of the PIK researchers, GFZ is seen as a scientific contributor to the climate crisis, with its strong (though decreasing) focus on natural resource extraction methods and technology (e.g. oil, minerals). On the other hand, PIK researchers were sometimes depicted as arrogant and excessively focused on the political positioning and public placement of its research insights. The cultural difference between the two institutes is driven more generally by a different focus regarding temporalities of the Earth System.

Figure 19: Guided tour in the Long Night of the Sciences in front of a PIK building on June 15, 2018. Source: Own photo

As a GFZ scientist declared during a guided tour (see Fig. 19) across Telegrafenberg, "we at GFZ are focusing on the past of the Earth. By contrast, the PIK is focusing on the future. Based on our findings about the past, they are developing computer models making statements about the future"[31]. While researchers from both institutes are using similar methods, GFZ is occupied much more with generating and analyzing empirical data, while the PIK uses these data in their models and simulations.

In fact, there have long been virtually no contact points between the researchers of the PIK and GFZ, despite of their physical proximity on Telegrafenberg and similar research objects. However, this situation is currently changing with the establishment of shared open data infrastructures (see Chapter V) concentrated at the science park library, which is establishing new points of contact between the four scientific institutes on the hill. Moreover, common challenges related to datafication and machine learning have also given rise to a shared new research network labelled *Geo.X Data Science.*[32]

Using open and closed doors

"It can be considered a paradigm of ethnography that failures in field access, averaging resistance and failure of attempts at understanding can also be used diagnostically, namely, as a method of relevance detection" (Hirschauer/Amann 1997: 19f) Even though my interlocutors were always interested in the concepts and activities of mediation, I was often confronted with boundary work (Gieryn 1983) between the world of scientific knowledge production and science communication. I noticed during my workshops and interviews that scientists are very sensitive and often disapprove certain boundary crossings between these alleged worlds of science (content) and communication (form). It is generally assumed that climate knowledge about the future is created within computer simulations; it is then carried to the public domain by

31 Guided tour in the Long Night of the Sciences 2018 on June 15, 2019.

32 https://www.geo-x.net/en/, retrieved on April 2, 2019.

project managers, visualizers and PR people. This way of maneuvering was especially apparent and routinized within the institute's so-called 'flagship projects,' which were accompanied by professionalized representation machinery.

It became apparent while doing my ethnographic research of scientific work at the PIK that scientists nowadays invest a lot of energy thinking about the representation and communication of their work outside scientific circles. Climate impact researchers are very concerned about the positioning of their knowledge outside their field. The very fact that I was invited as a visiting scientist at the PIK bares evidence of this. When I presented myself as a media scholar, I once received the answer: "Ah yes, we should do much more to communicate our knowledge to the media and the public."[33] The same researcher also told me that they integrated this module on communication and visualization into a project proposal for the *Joint Programme Initiative*[34] of the European Union, "because donors always want to have this now." Scientists generally rather feel obliged to fulfill this task to communicate with outsiders and see this as a priority of their work. They often seem to feel uncomfortable with this obligation as, for them, this is not their expertise. Having lunch with Jeremias Scholz and another young postdoc at the institute,[35] the former asked me what I think about such expectations about scientists communicating directly with the public. He mentioned that he was quite conservative in this regard. He sees the need to translate knowledge to broader audiences, but he also thinks that he and his colleagues do not dispose the necessary skills to do so.

---

33 E.g. discussion with Jeremias Scholz on September 9, 2019.

34 The Joint Programming Initiative "Connecting Climate Knowledge for Europe" is a pan-European intergovernmental initiative gathering European countries to jointly coordinate climate research and fund new transnational research initiatives that provide useful climate knowledge and services for post-COP21 Climate Action. See http://www.jpi-climate.eu.

35 Lunch with Jeremias Scholz and Layla Winston on September 9, 2019.

Several of my interview partners were overwhelmed by the impact that their excursions into mediation work had had in the public realm.

Compared to their peers in other academic disciplines, climate impact researchers can even be characterized as having been particularly concerned with the representation of their work outside the boundaries of their field due to the characteristics and peculiarities of climate impact research. While being abstract and complex, climate impact research is seen as highly relevant for other societal fields, such as policy, economics, health or security. Climate impact research is a relatively new scientific field. This means that there is a lot of space for experimentation and many aspects of the phenomena researched (climate, climate change and its implications for other systems) are not yet known. This high public exposure coupled with the instability of climate knowledge can be seen as an uncanny combination for the scientists at work. Be that as it may, climate researchers feel a strong need to communicate with the world outside their community of scientists. They do so engaging in multiple activities, depending on their occupational fields, professional skills and personality. These issues of 'communicating' or 'opening up' were often mentioned during my interviews with PIK researchers. At this point, it seems useful to mention a few aspects of my interview methods, as they structured the way in which my informants elaborated on forms of openness in their work.

Diagram elicitation

At the beginning of my fieldwork, I started a series of semi-structured interviews (Kvale 2007; Odendahl/Shaw 2001; Pierce 2008) addressing the issue of knowledge practices. The original idea was to trace the making of a 'prediction' and to investigate how such knowledge is transformed when it crosses boundaries of social worlds. This could, for example, be a prediction developed within a simulation modeling process at the PIK, which is then transformed to make it fit for the world of policy making. I had in mind Bruno Latour's cascade of inscriptions tracing botanic and pedological knowledge practices observed within fieldwork in the Brazilian Amazon (1999a) as an underlying theoretical concept. In contrast to Latour's example, it became apparent that scientific practice in simulation modeling could not be observed 'in action' by just being there with the researchers. This would have meant staring for months at the back of modelers, staring, in turn, into their computer screens. Writing computer code as the primary daily practice in simulation modeling is ungraspable by traditional ethnographic means of observing, describing and/or recording embodied movements and articulation. As a matter of fact, this might be the case for the observation of many other working practices within today's knowledge society. The second problem is the distributiveness of activities in simulation modeling within space and time. Climate impact modelers may work on three or more projects in parallel. Realizations (i.e. calculations) of simulation models take time (days, weeks). Therefore, modelers would start one calculation in the supercomputer, continue to work on the computer code of a second model, go to meetings of yet another project, then check in again on project one and evaluate the results of the simulations. In so doing, they work together with others who may not be present on the same floor, in the same city or country. In order to deal with these methodological issues, I decided to conduct interviews with simulation modelers and trace knowledge practices *ex post*. After some exploratory discussions, it became apparent that it would be impossible to understand and retrace knowledge practices just by letting scientists talk about their work. The only way to stabilize such conversations was to make use of visual artifacts and structure the interviews around these representations of scientific work. Drawing on Douglas Harper's approach of photo elicitation (2002), I asked every interviewee to bring a visualization (mostly diagrams or maps) that could be used to explain some elements of their current or past working activities. I spent a considerable time with some of my interviewees looking at their visualizations. Torsten Casius, for example, could explain to me his work on developing and running the 'SWIM'[36] model simulating European waterways.

As he told me, the most considerable challenge was to find, collect and harmonize the datasets representing spatiotemporal values for hydrological and land use-related information. On the one hand, it would necessitate calling administrators or researchers in the Czech Republic, Poland or the Ukraine and asking them kindly to donate their datasets for the research carried out by the Potsdam Institute. If phone calls do not suffice, he would have to travel there to collect the data for himself. In this sense, 'collecting data' does not mean getting in contact with the simulated phenomenon in question (e.g. the Donau river) but with the actors holding back the datasets on their computers.

Figure 20: The working environment of Torsten Casius (visualized datasets on land use for SWIM simulation in GRASS software). Source: Own photo

36 Soil and Water Integrated Model.

As we can see in this example, climate impact simulation is not only a 'technical' challenge. The international character of the phenomena simulated also poses considerable social and political challenges, drawing together all necessary ingredients. As we will see in Chapter V, the movement around open data aims at reducing the barriers to receive, access and use these ingredients. Torsten Casius also talked about issues of representation in transformative data practices:

> You often have the trouble to transfer the whole thing to SWIM's own data formats. SWIM takes soil profile files according to a completely defined scheme. These are ASCII files. And, of course, this format is not provided with the input data you have. You have some kind of ACCESS database with soil parameters in it, but that's not the format that SWIM needs. So, this transformation creates a lot of work. That's why colleagues developed methods to convert the data from the international world map – very extensive Python or R scripts. [...] Well, I had a bit of a bad feeling when I processed the data with that. [...] The scripts threw out two or three soils without parameterization, which I had to puzzle out by hand [...]. And you never know: If it is so complex; is it correct then? Doing everything by hand, however, is not less error-prone. And it's definitely a lot of work. (Interview Casius, translated by the author)

As we can see, the establishment of circulating references between different inscriptions in a research process is as relevant in 'digital pedology' (i.e. manipulation of land use data) as in its analog version described by Bruno Latour: "The transformation at each step of the reference [...] may be pictured as a trade-off between what is gained (amplification) and what is lost (reduction) at each information-producing step" (1999a: 71).

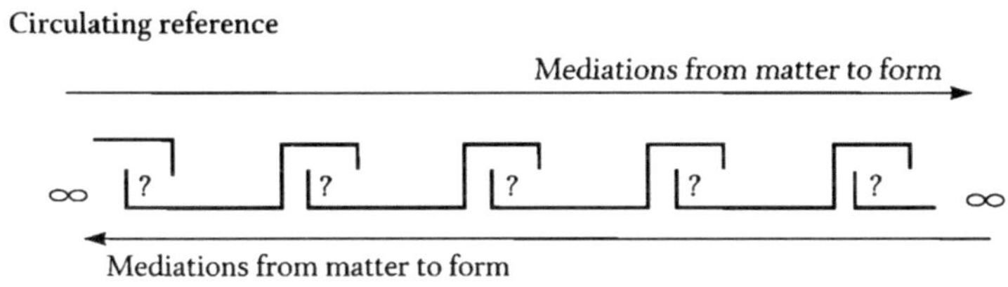

Figure 21: Circulating reference. Source: Latour (1999a: 73)

I started to develop visual guidelines to structure my interviews with informants as a second methodological device (see Fig. 22). In the middle of the map is a square, where my interview partners could describe the project that would be discussed within the interview.

Figure 22: My visual interview guideline. Source: Own visualization and photo

The left side of the map is meant for technical inputs to the computer model, such as algorithmic code and datasets. On the right side, is space for outputs (e.g. data, code, visualizations). However, these elements are complemented by other ingredients, such as human contributors, financial means, essential skills of the actors involved and infrastructure required for the project. In my interviews, I first asked my informants to talk about and fill out the input-output elements in the visual guideline. The aim here was to make them comfortable and speak in their (technical) language about things they know best. Later, I focused more on questions that are normally not covered within the daily in-house discussions at the institute; for example, the description of tacit knowledge (MacKenzie/Spinardi 1995) that make up a good modeler or problems with the 'delivery' of more comprehensive outcomes (e.g. lacking success to sustain project results and make them useful for other scientists). At the bottom of the page, the guideline features a visual representation of Bruno Latour's chain of circulating references (1988). This underlying heuristic of the interviews was rarely addressed explicitly, but sometimes my interviewees would ask about the 'theory' behind my research.

The right timing for openness

In the interviews, researchers would often mention communication activities, such as 'public and press relations,' 'policy advice,' 'stakeholder engagement,' 'outreach,' 'science communication,' 'open sourcing,' 'science education' and "'open science.' At the beginning of my field research, I interpreted the labels of the interaction between science and nonscience as variations or even synonyms of the same practice. The discussions with scientists were mainly driven by the question 'how to communicate information to different audiences,' so-called 'user groups' or 'target audiences.'

I had to reevaluate this interpretation significantly after a year working at the PIK. It became clear that the primary issue of controversy was not how to talk to whom but *when* to talk. All my interview partners were equally interested and ready to pass on their scientific insights to others, including those outside their WGs, their institution and the scientific community. By contrast, there have been very contradictory views among interviewees whether such openness should involve elements other than stabilized scientific evidence and when such elements should be passed on.

What we may call the 'conservative view' of timely openness would suggest that the only task of researchers is to conduct scientific experiments. If these experiments are successful, the scientist may write an academic publication communicating his/her insights. Accordingly, he/she may also engage in other activities of science communication. In the best case, this task of science communication is delegated to professionals, such as the PR department, a media agency or perhaps even another dedicated academic institution. An example of the latter is a collaboration between the Potsdam Institute and the design department of the School for Applied Sciences in Potsdam (FHP). The FHP interaction designers created *A Brief History of CO2 Emissions,* a dynamic animation featuring insights from the PIK's simulation of the SSPs. While the animation was composed using the newest available design components by FHP, it mediated highly stabilized and published scientific knowledge.

On the other hand, we can identify a 'progressive view' of timely openness, which operates with terms such as 'open science,' 'open data,' 'stakeholder involvement,' or 'participatory design.' We may focus on *Open Science* as the most fashionable umbrella term, aiming at the incorporation of all these other approaches. As Benedikt Fecher and Sascha Friesike highlight:

> 'Open Science' is one of the buzzwords of the scientific community. Moreover, it is accompanied by a vivid discourse that apparently encompasses any kind of change in relation to the future of scientific knowledge creation and dissemination; a discourse whose lowest common denominator is perhaps that science in the near future somehow needs to open up more. (2014: 11)

Fecher and Friesike identify a number of different schools of thought in the understanding of open science: The Public School, the Infrastructure School, the Pragmatic School, the Democratic School and the Measurement School.

A lot of scientists believe, for example, that science must be accessible to the public ('Public School'). Consequently, they engage in concrete communicative activities, such as science blogging, science PR and experimentation with citizen science. Other scientists are more concerned with the legal and financial restrictions to knowledge dissemination and believe that knowledge must be freely available ('Democratic School'). The Open Access Publishing movement is a prime example of these concerns. In another reading, Open Science is more about making collaboration within the research community more efficient and effective ('Pragmatist School'). This may embrace a valuation of digital communities and collaboration platforms, and expected network effects. Another variation of open science is concerned mostly about the aspect of evaluation ('Measurement School'). Many researchers consider the current dominant evaluation systems, such as impact factor schemes, as problematic. They propose alternative possibilities of evaluation, such as altmetrics and open peer review. Finally, an Open Science theme of increasing prominence is infrastructure ('Infrastructure School'): "The infrastructure school is concerned with the technical infrastructure that enables emerging research practices on the Internet, for the most part software tools and applications, as well as computing networks" (ibid.: 36).

The proponents of the progressive view of timely openness can be attributed mostly to the infrastructure school of Open Science. These actors experiment with ways to make the scientific process itself open from the beginning. Equally, they engage in futurework, aiming at a design of research elements which enables the reuse by others. In the following chapters, I will address three ways to 'open up' research elements at different points in time of the scientific process: Visualizations (Chapter III), software (Chapter IV) and datasets (Chapter V). In Chapter VI, I will discuss an example beyond science which makes use of multiple 're-usable elements.'

III. Future images

As we will see throughout the chapters to come, scientists employ different strategies to make their science 'more open.' In the following, we will address one particular project of mediating simulation modeling knowledge to communities beyond the world of science: The development of the map-based online portal CIO. Very fortunately, I had the opportunity to become part of the CIO project team for the period of one year. This position did not only give me more privileged access to scientific practice, it also provided an opportunity to experiment with different methods of analyzing and characterizing geomedia platforms and their user communities.

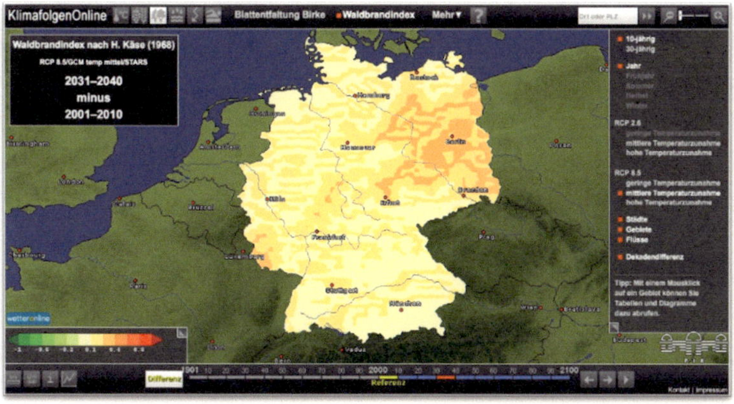

Figure 23: Snapshot of the CIO Graphical User Interface (GUI), as accessible through www.climateimpactsonline.com. Source: CIO Website

The experiment was guided by the question, "How do actors cope with the challenge of 'opening up' scientific knowledge, and how do they engage in parallel tactics of stabilization within this opening process?"

The CIO has been an experiment for PIK in multiple ways. It aims at translating numerical simulation outputs to a format that should be depictable on the web and understandable for nonscientists. The underlying scientific data of the platform builds on a number of research projects at PIK, calculating probable future impacts of climate change in Germany by means of computer simulations. The simulations were essentially driven by the regional climate model (RCM) developed at PIK (Orlowski 2007). Building on the highly detailed spatiotemporal temperature data generated by STARS, further impacts of these temperature changes were simulated by climate impact models such as SWIM (hydrology), IRMA (agriculture) and FORESEE (forestry). Traditionally, the way to make scientific knowledge public generated by these simulation processes is to write an academic publication. The insights from water-related simulations, for example, are gathered in the anthology *The Elbe River in Times of Global Change: An Integrative Assessment*[37] (Wechsung et al. 2014). The cover of the large-formatted book depicts a puzzle featuring a photographic image of the bed of the River Elbe in front of the skyline of the east German city. Within the analogy of the puzzle, the book aims at drawing together a complete image of the impacts of climate change expected in the Elbe area in Germany. The publication entails methodological explanations of the simulation experiments and interpretations of the results.

The transdisciplinary team of scientists also published their insights in a more accessible format – the *Elbe Atlas* (Wechsung et al. 2011). The atlas depicts a variety of maps showing multiple impacts of climate change on rivers, landscapes, industries and cities. While the

37 Original title in German: "Die Elbe im globalen Wandel: Eine integrative Betrachtung."

development of the atlas has been a cumbersome work for modelers and cartographers at the PIK, it did not change or interfere with the process of scientific knowledge production essentially. As a matter of fact, the acts of 'opening up' knowledge beyond the scientific community were undertaken after all simulation experiments had been carried out and conclusions had been made. The maps of the atlas illustrate the stabilized insights of the scientific work. From the perspective of the scientific process, the atlas comes after the publication, similar to a press release.

The CIO platform differs from these traditional models of science communication. It was originally developed to inform decision-makers, administrators and scientists about the local impacts of climate change in Germany. Subsequently, the portal was adapted to serve other user communities, such as pupils in German schools. Secondly, it tried to operationalize the concept of *climate services* (Krauss/von Storch 2012; Vaughan/Dessai 2014), enabling the reuse of climate data by commercial and noncommercial actors. Accordingly, the construction of the platform had been funded by the European Institute of Innovation & Technology and its climate-technology stream Climate-KIC. It was set up as a public-private partnership between PIK and the commercial weather-forecasting company WetterOnline. Thirdly, the platform was probed as an education tool to bring the topic of climate change into German schools. This alteration of the user community brought a number of fundamental issues to the surface regarding the aptitude of maps and diagrams to mediate a multifaceted phenomenon such as climate change.

Imagining users

As we have seen before, designing technologies and infrastructures always entails imaginations about prospective audiences and users. This is not only true for physical technologies and infrastructures but especially for those in the sphere of the digital, as Sally Wyatt highlights in her analysis of users and nonusers of the internet:

> To many people, cars reflect wealth, power, virility, and freedom. The Internet promises many of the same attributes on an even larger scale, with its possibility of global reach. The symbolic value of having Internet access is often presented as a sign of inclusion in a high-technology future. (2005: 70).

These promises might not be part of a homogenous master narrative (Star 1999) but be heterogeneous and sometimes contradictory. Nevertheless, drawing relationships between concrete technological devices and popular imaginaries around technology use can be helpful to understand how these devices work. A prominent imaginary for environmental geobrowsers, such as CIO has certainly been the 'digital earth,' most famously described by former US Vice President Al Gore as early as 1998. Gore highlights that:

> A new wave of technological innovation is allowing us to capture, store, process and display an unprecedented amount of information about our planet and a wide variety of environmental and cultural phenomena. (1998: 89)

Building on this availability of technology and information, he then imagined a new kind of interfacial device to connect people with such information.

> The Digital Earth would be composed of both the "user interface" – a browsable, 3D version of the planet available at various levels of resolution, a rapidly growing universe of networked geospatial information, and the mechanisms for integrating and displaying information from multiple sources. (ibid.: 91)

Interestingly, Gore provided a very detailed description of a prospective user and a concrete situation of interaction:

> Imagine, for example, a young child going to a Digital Earth exhibit at a local museum. After donning a head-mounted display, she sees Earth as it appears from space. Using a data glove, she zooms in, using higher and higher levels of resolution, to see continents, then regions, countries, cities, and finally individual houses, trees, and other natural and man-made objects. Having found an area of the planet she is interested in exploring, she takes the equivalent of a "magic carpet ride" through a 3-D visualization of the terrain. Of course, terrain is only one of the many kinds of data with which she can interact. Using the systems' voice recognition capabilities, she is able to request information on land cover, distribution of plant and animal species, real-time weather, roads, political boundaries, and population. (ibid.: 89)

As media scholar Pablo Abend has shown, the most direct interpretation of the Digital Earth concept has been the software Google Earth, which mainstreamed geobrowsing as a popular (i.e. widespread) practice (Abend 2013). More than ten years after the launch of Google Earth, it may be debated whether geobrowsing can be characterized as a genuine media practice or rather a short-term spectacle around the astonishing aesthetics of a new product. Nevertheless, Google Earth and similar devices have spearheaded aesthetics, functionalities and practice which have now been stabilized around all sorts of digital mapping tools, including route navigation devices. In addition to these popular practices, Digital Earth has been a pulse generator for more techno-scientific software, such as ArcGIS and Google Earth Engine, as well as for online data explorers built in the context of science communication (Hewitson et al. 2017; Neset et al. 2016).

CIO is an example of such a data explorer. Its brochure claims that the internet portal "enables you to investigate the impacts of climate change on Germany with just your computer" (PIK 2012). Such a characterization evokes great expectations: To provide a direct interface between people (users) and a complex body of knowledge (impacts of climate change in Germany). They come with an understanding of design aiming at a transparency of the interface (Bolter/Gromala 2003: 4): An interface is well-designed if it remains invisible to the user and establishes immediate (not mediated) access to the relevant information. However, there is a second aspect to transparency in the description. The platform aims at making climate information independent from the scientist, therefore, lifting limitations of information transfer in time and space. People will not be forced to visit the scientist and vice versa to learn about local climate change impacts. The information is always available and can be accessed everywhere though the web:

> The sole precondition for using the internet portal is an up-to-date internet browser. Anyone can use the portal. Administrative or technical hurdles have been dispensed with (registration or plugins for the browser are not required). The information is available to users free of charge. The stated aim is to win over the largest possible number of users – it is envisaged that all members of the general public who are interested can access it. (PIK 2012)

Semiotic elements

The first experiment in investigating CIO was to engage in a semiotic analysis of its visual components – maps, diagrams, dashboard elements and text. This analysis can draw from analytical approaches in visual studies (Mitchell 1987; Müller 2011; Panofsky 1972; Rose 2001), the description of scientific and technical imagery (Bredekamp et al. 2012) and iconological analysis of cartographic representations (Harley 1988).

The visual appearance of the Graphical User Interface (GUI) of CIO mashes up the aesthetics of weather maps, online geobrowsers (e.g. Google Maps) and scientific tools for visual analytics. A geographic map centered on the territory of Germany occupies most of the space available in the browser window. A number of navigation elements are structured around the map enabling the user to browse a variety of climate-related parameters for different time scales, geographic locations and alternative scenarios. On the top left, a number of buttons (or 'icons') represent different sectors of climate change impacts, namely, climate (thermometer), agriculture (cereals), forestry (trees), hydrology (wave and water drops), energy (lightning) and miscellaneous (person swimming in water). Users can then choose between different climate variables, such as mean temperature, wildfire risk or number of swimming days per year, by clicking on the different sectors. The information on these variables is represented on the map in false color, similar to a weather map. A slider at the bottom of the interface enables users to navigate a time axis. The timeline starts on the left with the year 1901 and ends on the right with the year 2100.

Users can choose between absolute or relative values when navigating the timeline. Absolute values, for example, could display the mean temperature in the decade between 2070 and 2080, while clicking on 'difference' would show the change of variables between two decades. More levels of complexity are introduced in a text box on the right, where users can change between different averaging intervals (thirty or ten years and seasons or annual), different scenarios of the future and orientation aids to be displayed or hidden in the map (cities, areas, rivers).

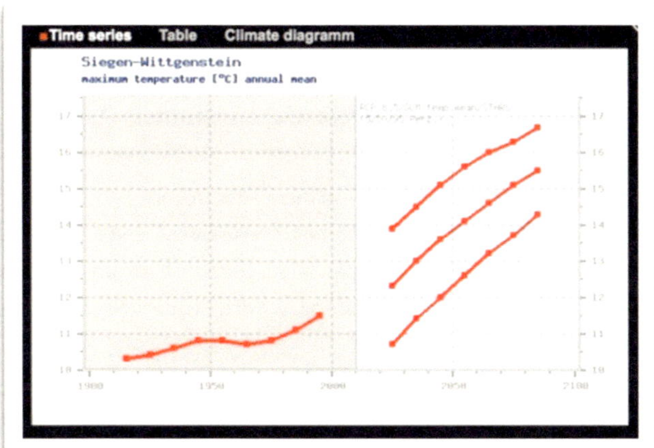

Figure 24: Maximum temperature (°C) annual mean in the region of Siegen-Wittgenstein with three different variations of future changes (model uncertainty). Source: climateimpactsonline.com

Users can also zoom in to inspect climate variables within a specific region on the map itself. A pop-up window opens by double clicking on a location and shows the local data in a histogram.

The fairly sophisticated navigation of the portal has been a frequent matter of debate within the project team. It exemplifies the challenges in translating scientific knowledge generated within computer simulations to make it accessible to a broader audience.

Unfolding the making of maps

In order to understand these challenges, we have to dig deeper and engage in an unfolding of the sociotechnical relationships within the construction of the mapping (Kitchin et al. 2013). Technically, CIO is a comprehensive Geographic Information System (GIS), hosted on the server and operated through the weather forecaster WetterOnline headquartered in Cologne. The primary challenge of CIO from a technical perspective was to translate modeling outputs of the STARS simulations into a format that could be depicted by the maps and GIS of WetterOnline. I interviewed several of the CIO project team members to learn more about this process.

Clemens Rechstein told me that the output data of all the relevant simulations at the PIK had to be collected and put into a homogenous matrix stored on the central servers of the institute:

> We have a giant variable list. It has these subcategories. There's the variable name, a variable description, a unit of measurement, and a few other things. They're all in there. And it's expandable, too. It's basically a matrix, where all individual data are stored. (Interview with Clemens Rechstein, translated by the author)

The format was optimized to fit the affordances of the GIS of the company WetterOnline, which is normally used to depict weather maps for Germany.

> And then we agreed on a format with them, actually a relatively simple format, the maps with a certain resolution. (...) At any location, with a certain precipitation, tree growth and so on. (Interview with Clemens Rechstein, translated by the author)

Figure 25: Clemens R. pointing at 'data' of the CIO.
Source: Own photo

Such data curation has multiple facets. Data have to be aggregated and restructured to ensure performance and interaction in the GIS. While the original datasets mostly have a European scope, data in CIO needed to be 'cut out' to depict only the territory of Germany. In addition, some simulation models do not 'respect' political borders at all but run within other logical entities (e.g. cross-national river flows).

Figure 26: Internal (right) and published version of CIO.
Source: Own photo

Rechstein developed his own local visualization tool that runs on the PIK's servers to do the data cleaning work and enable communication within the relevant group of data contributors. Figure 26 shows this internal version (monitor on the right) and the public version provided by WetterOnline (laptop on the left). As one can imagine, these tasks are time-consuming and so are maintenance and updating works. As Clemens Rechstein tells me, these activities are typically not accounted for in scientific project funding. Accordingly, he had to find time for these tasks between his other projects.

After receiving the data from PIK via an application programming interface (API), WetterOnline then generates images out of the data:

> And they created these pictures out of it. Everything you see here is practically not calculated on the fly, but calculated in advance. Every picture. And there are thousands of these pictures. If you zoom in here now [zooms in], it's a new image. And that's what they did at WetterOnline. They calculated it all in advance. (Interview with Clemens Rechstein, translated by the author)

The choice of color schemes has been one of the major issues of debate within the project team. Researchers are often reluctant to break rules of coloration in popular representations of their work because they fear giving away scientific integrity. A similar debate is linked to interpolation. In fact, the final images do not depict the original data points but use a smoothening algorithm which makes them visually more appealing:

> Well, you don't see the pixels here. You would see them theoretically. You can't see them here, however, because WetterOnline smoothed everything. (Interview with Clemens Rechstein, translated by the author)

The interpolation creates the feel of the weather map and draws the aesthetics away from scientific visualization. The resolution of the data in particular cases has also been lowered for legal concerns, for example, in the depiction of agriculture:

> For agriculture, they [the relevant lead scientists] did not want to have the data maximally resolved, only to federal state levels. It's a matter of data protection that you can't zoom onto the level of individual farms. Agricultural yields are subject to data protection. (Interview with Clemens Rechstein, translated by the author)

After the rendering of the images, the latter can be integrated into WetterOnline's GIS, which finally enables the navigation and depiction in the browser:

> [...] and then you can navigate here. The whole navigation has been programmed by them [WetterOnline]. (Interview with Clemens Rechstein, translated by the author)

As we can see with all these small but manifold transformations, the information depicted in such platforms is all but 'raw scientific data.'

Paradoxically, a lot of translation work is necessary to make information appear as 'raw data' which can be navigated interactively in a web browser.

While the PIK researchers have always highlighted a hierarchy in the division of labor between them (scientific content) and their project partner WetterOnline (technical operator), this boundary work between science and technology or content and form did not always match the actual balance of forces in the project. Along the line, the WetterOnline GIS had been the stable (we may even say rigid) central element to which all other more fluid (human and nonhuman) elements had to position themselves. This includes elements such as the GUI's symbols, the scientists and the datasets. It is important to highlight here that this is not about traditional principal-agent problems (Jensen/Meckling 1976) between institutions (PIK, WetterOnline) but about a stickiness of (digital) infrastructure.

The public side of the interface

What happens if data mediated through such systems and interfaces go public? How do people make sense of visualized data in concrete settings of mediation and translation? We can draw on approaches from the fields of symbolic interactionism and praxeological media research to analyze situations of open science in action. Karin Knorr-Cetina argues in an article published in 2009 for a renovation of Goffmanian thinking to deal with situations including "synthetic components" (2009: 63), understood as elements mediated through electronic information technologies. In the traditional understanding of symbolic interactionism, a situation was "a physical setting or place with a physical coming together, a human encounter, typically taking place" (ibid.: 64). In the context of a situation, there was something "analytically prior and theoretically foundational about physical encounters in physical settings" (ibid.). However, in a networked society, many areas of everyday life have migrated to the "internet" or "virtual spaces" (ibid.:

65). For Knorr-Cetina, situational analysis needs to conceptualize "the presence of different electronic media and their contributions to both 'situations' and the coordination of interaction" (ibid.). Synthetic elements may change the temporalities of the situation, thus, restructuring the interaction order: "[S]ynthetic situation's assemblage and projection is a continuous project" (ibid.: 71) and "behavioral settings may extend in space and time" (ibid.: 64). Knorr-Cetina states that synthetic situations "carry a time index; their components tend to require frequent or continuous updating or else their iterated presentation as still 'live' and relevant" (ibid.: 72). Of course, spatial concepts never purely denied temporal processes, but they tend to treat them as externalities:

> [T]hey imply that time is something that passes in the spatial environment and is extraneous to the environment itself. Presumably, we also express durability through spatial concepts. The synthetic situation, however, is inherently in flux; it has none of the durability of a physical situation. (ibid.: 73)

Knorr-Cetina highlights three features of synthetic situations in her conceptualization: 1) "They are entirely informational, 2) ontologically fluid and 3) may project a party to the interaction" (ibid.: 70). Empirically, Knorr-Cetina's conceptualization of synthetic situations builds mainly on observations of working arrangements and their *scopic systems* in the field of high-frequency trading:

> When combined with a prefix, a scope (derived from the Greek *scopein*, "to see") is an instrument for seeing or observing, as in periscope. In such markets, a scopic system is an arrangement of hardware, software, and human feeds that together function like a scope: like a mechanism of observation and projection, here collecting, augmenting, and transmitting the reality of the markets, their internal environments and external context. Within this domain, the mechanism is reflexive: the system mirrors a world that participants confront like an external reality while also being part of it and contributing to it through their postings and transactions. (ibid.)

The synthetic situation described in the following is certainly not as interactive as the one described by Knorr-Cetina. Nevertheless, her conceptualization of the synthetic situation may serve as a methodological device to engage in further analysis of the socio-technical mediations between (materialized) data and (becoming) users. The following ethnographic vignette describes a concrete operationalization of a data-interface 'in the wild.'

Every year, about seventy scientific institutions in Berlin and Potsdam open their doors and invite the so-called 'interested public' to the Long Night of the Sciences, including numerous presentations, experiments, meeting spaces and other formats of science communication. Between other projects, PIK presents the educational version of the web portal CIO at the event, which is advertised in the program of the Long Night with the following words:

> Climate impacts in Germany. What does global warming mean for the individual regions in Germany? Where does agriculture have to adapt, where can we go swimming more often in the future? KlimafolgenOnline provides answers. Demonstration, information stand: From 17.00 to 23.00, rotunda, ground floor. (Verein Lange Nacht der Wissenschaften e. V. 2017)

I receive a staff badge for the Long Night of the Sciences 2017 thanks to my status as a visiting scientist at PIK. While not having any fixed obligations at the event, I promised to help out wherever needed. Together with other visitors (mostly families, pensioners and scientists), I reach the Telegrafenberg at 6 pm, around the beginning of the Long Night. When I arrive on the mountain, the event is already in full swing; many visitors have made it to Potsdam and up the mountain. Events take place at PIK as well as all the other research institutions at the Albert Einstein Science Park (GFZ, AWI and AIP).

The main PIK building hosts a number of exhibits, mainly posters, experimental installations and computer screens. A scientist is assigned to every exhibit, explaining his/her work to visitors by making use of the artifacts available. The setting for CIO is the following: A large television display on a stand shows the GUI of the new CIO educational version (beta) in the full screen mode of a web browser. As a backup, a second tab enables one to open the stable classic version of the portal. A bar table is set up in front of the screen, featuring a mouse for GUI navigation purposes and a stack of well-designed brochures introducing the web portal. When I arrive, Sabina, the intern of the department, is standing in front of the screen at the bar table. Sabina is studying Global Change Management at the University of Eberswalde, where she is writing her master thesis on the deployment of CIOs in schools. After a while, she is replaced by the educational expert, Irina, who is leading the educational project, which aims at explaining climate impacts in Germany using the web portal. Numerous visitors come by and interact with the CIO setting - the scientists, the stand, the visual interface. Two older ladies, for example, approach the stand and ask Irina what this is all about. Irina explains that the screen visualizes the climate consequences in Germany. The two women are particularly interested in the risk of wildfires and the drought in Brandenburg, as they were reading about this in a newspaper. They ask Irina how this might develop in the future. Irina chooses the visualization of wildfire risk on the CIO map and explains the connection between a lack of precipitation, flat-rooted pines and the danger of desertification in parts of Brandenburg.

Figure 27: Interactive setting of CIO at the Long Night of the Sciences 2017. Source: Verein Lange Nacht der Wissenschaften e. V. (2017)

Half an hour later, Irina hands the stand over to Tim, who significantly developed the web portal and oversees the scientific soundness of the climate information shown. After a while, a visitor, about seventy years old, reaches our stand. He asks: "What is this about?" and lets Tim introduce the functionalities and information shown in the CIO portal. The man listens. After Tim's introduction, the visitor mentions having read about a connection between climate change and the Syrian war. According to the relevant article, climate change had been one of the triggers of the war. Tim mentions that there are different views on this relationship. In particular, he cites a PIK study which had shown that this connection was not as strong as the media sometimes reported. According to the PIK study, the famine causing social unrest in Syria before the civil war had been triggered by high food prices caused by the sanctions against Russia, not by the drought caused by the climate. The visitor thanks him for this explanation and continues on his way to the next stand. The next visitor seems to have a background in the natural sciences. He wants to know from Tim how the different variables shown in CIO are related.

I move over to take a look at the other exhibits in the context of my ethnographic endeavor. Later, I enter the room next to the cupola to meet some of my other informants. Frederik Willkomm is standing behind a counter selling local wine to visitors. Both he and his father were working at the vineyard based in the Brandenburg region. When I approach him, he is just chatting with a visitor: "Yes, there might indeed also be rare cases where climate change is doing good. For example, for wine growing in Germany. The production of this regional wine will be much easier with growing temperatures." I buy a glass of 'Solaris' (a German crossbreed grape grown here in Brandenburg) and go back to the CIO stand. When Tim is called to another event at short notice, I take over the mediation myself. When I am standing in front of the portal, a middle-aged father and his primary school-aged daughter join me. The father asks what this is all about. Since I had just looked at the Huglin index for wine-growing, I told him that the consequences of climate change on winegrowing in Germany could be seen here. The two of them listen to me and the father asks how the scale works. I explain that these are different types of wine that could be cultivated. The two do not understand at first what the navigation on the portal shows or what exactly can be seen. When I explain that you can switch between different climate scenarios with a button on the right, the father says, "Ah, now it's clear." The girl says that climate change could lead to a situation where there are no longer enough bees available to pollinate the plants. The father then asks whether there is any information about the danger of forest fires in the Osnabrück area. He owned a piece of forest there. We look at the consequences of the climate change for the forest fire danger and the forest formation for beech and birch, which grow in his forest. The father seems very interested in this information. After one hour, I hand the stand over to Tim, who arrives with his own family. I continue my visit to the other exhibits, leave the Telegrafenberg at around 11 pm and take the train back to Berlin.

Drawing from Adele Clarke's (2003) situational mapping approaches, we may visualize the situational setting at the Long Night of the Sciences, as shown in Figure 28. The components of the synthetic situation include human elements (visitors, scientists, science communicators), physical elements (a screen, a table, a computer mouse, brochures, computer servers, cables, architecture) and symbolic elements (the images and text shown within CIO and represented within the accompanying brochure about CIO).[38] We can identify and characterize interactions between the constituting elements of the situation by means of a relational analysis. Every interaction establishes a relationship, which can then be visualized within the situational map. A scientist, for example, establishes a connection with a visitor. The scientist uses the mouse to interact with CIO. The scientist illustrates an argument by pointing to the maps depicted within the platform. There is also an additional category of elements, such as the internet connection, the GIS, the science park and the science communication event. These components and their relationships differ from the others regarding their temporal permanence. While relationships between the scientists, visitors and the GUI emerge within the temporal period observed, others were stable throughout the period of five hours. We could differentiate these two classes of relationships – emerging and stable – through the dichotomy between agents and (infra-)structure, but, as we see through this study, this distinction depends on our spatial and temporal scope of analysis, establishing different constellations of relationships. In our case, relationships that emerged before the temporal window of the Long Night of the Sciences observed became 'structural' and elements emergent within the observational period became 'agents.' There is no

38 The categorization into human, physical and symbolic elements is derived from Rammert (2007: 19).

substantial distinction between these elements, for example, regarding the dichotomy of enduring stuff and mobile people. Considering this relative status of all entities and relationships, I prefer the term 'element' to other categorizations, such as 'agents,' 'actors'[39] or 'actants.' The term 'element' highlights its fluid ontological status and is agnostic to its role as 'actor' in ever-changing actor-networks.

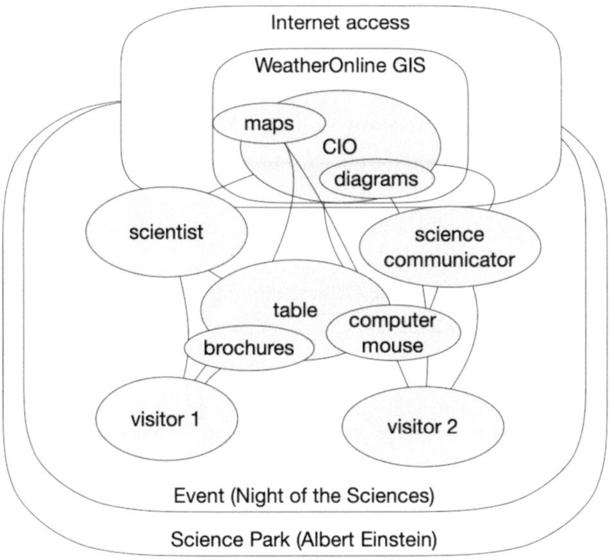

Figure 28: Interactional space at the Long Night of Sciences 2017. Source: My own visualization

The same agnosticism can be applied to the visual representation of the situation, be it a photographic image (Fig. 27) or a situational map (Fig. 28). There is no claim of substantial truth in these visualizations. They have to be considered more as perspectives by us or others, mediated through different cultural techniques of inscription (e.g. mapping,

39 I will use the term 'actor' occasionally to refer to human participants of a situation or technological setting.

photography). It is useful to overlay different genres of inscriptions and engage in techniques of triangulation, as proposed in Chapter I. The result is a collage (Kalthoff 2010) which highlights different perspectives and readings of a situation without raising claims for substantial representation.

Script of a situation

While there have been varieties and iterations, interactions with CIO at the Long Night of the Sciences all ran according to a similar interaction order. Inspired by the program to 'run a kitchen' (compare Fig. 12), I developed a script formalizing the interactions regarding the Long Night of the Sciences:

(1) Navigating through the architecture of the PIK building and the suggested route of the exhibition, visitors approach the setting of scientist, stand and screen.

(2) An exchange of greetings takes place between the scientist and the visitor(s).

(3) The screen catches the attention of the visitors.

(4) Visitors are asking: "What is this about?" or "Can you tell us something about this?" Alternatively, the scientists propose, "May I tell you something about it?"

(5) The scientist chooses one topic addressed within CIO, such as forestry, water or agriculture. Making use of the map, he/she explains what impacts are observed or expected according to the research at PIK.

(6) The visitors take the scientist's story as a trigger for a broader discussion of issues, which are not directly related to the information shown on the map of CIO.

Analyzing this sequence and comparing ('overlaying') my observations ('visualizations of observations') with those made in other situations in the field, one issue is increasingly salient: Most alleged 'users' of the platform were reluctant to touch the input device (mouse) suggested, navigate with the dashboard (pointer, icons, menus, map, sliders) proposed or browse through the data and curated information on their own. The fact that no visitor approached the CIO setting (screen, table) when the human hosts were absent confirms this assessment. In this sense, the GUI on the screen without the scientist was like a telescope without a scientist. It may be fun for children and curious adults to gaze through the lenses of a telescope, but this uninformed interaction with a scientific instrument is not likely to increase the spectators' understanding of stars. By contrast, it may make the spectator interested in acquiring this information by seeking an interaction with a scientist or other informational mediator (e.g. climate information websites, Wikipedia, online press dossiers on climate change).

To put this differently, there is no established practice of 'using' a technological artifact like the CIO platform by browsing through datasets or, more realistically speaking, through maps and diagrams. By contrast, the CIO has been extraordinarily useful as a presentation tool for researchers explaining local climate change impacts to nonexpert audiences. In this traditional format of science communication, the researcher talks about the insights from his/her experiments and puts them up for debate. In a contemporary understanding of science communication, this goes beyond reducing the 'information deficit' of the audience but may include debates about perceptions and consequences at an eye level (Sturgis/Allum 2004). This discussion (see 6 above) can be considered as the essential part of the discussions, considering its temporal length, density and commitment among participants. The topics of discussion witnessed included: What impacts are expected at my home or living area? What changes in my social world, for example, family, children, field of work? As an expert, how would you rate the facts XY discussed in the media (e.g. climate as a trigger for social conflict in Syria, climate change as a threat to bee populations)? What is your personal opinion regarding the controversy about the robustness of climate science and model-based predictions? Arguably surprising for the climate scientists, concerns about the data depicted in the maps were not among the questions posed most frequently.

In the context of science communication in the fields of the natural sciences and climate science in particular, the preferential artifacts facilitating such debates are diagrams and maps (Schneider/Nocke 2014). I could observe on various occasions that maps and diagrams are essential elements enabling climate impact researchers to communicate the findings of simulation experiments to nonexpert communities. As a matter of fact, diagrams and maps support the credibility of the scientist, as they establish a circulating reference between the researcher present and the scientific process, the data and the other scientists involved. Visualizations materially realize scientific data in a situation (also see Chapter VI). In the words of Bruno Latour, they mobilize allies in the situation, preventing the scientist from being confronted alone with an audience. The presence of visualizations in settings of science communication are essentially anchoring the discussion and building trust.

This absence of proper 'users' and routinized 'user practices' poses a methodological challenge for investigation. In fact, I experimented with multiple ways of identifying and characterizing 'user' practices in CIO. This includes interviews with alleged power users (experts, local decision-makers), observing 'use' by looking over people's shoulders and recording interactions with digital tools (Abend et al. 2012). However, it became obvious that these experimental settings would create new users instead of describing existent practice. In-depth ethnography was the only way forward here to grasp the fragile, constantly evolving media practices involving the CIO platform.

Educational work

In 2015, the German Federal Foundation for the Environment (DBU[40]) funded a project by PIK called PIKee, which aimed at experimenting with online-based environmental education. The idea was to build on the experiences with the CIO platform to bring the topic of local climate impacts into German classrooms. The existing platform should be adapted within this perspective to fit a particular user community: Teachers and pupils. The work should be implemented by a team of pedagogic experts, together with the PIK scientists in charge of the CIO platform. The project activities started with a series of (about 40) workshops conducted with teachers and pupils in different schools throughout Germany. On the one hand, the workshops were conducted as a participatory design process, aiming at a technical reworking of the existing portal to fit the needs of a new audience. On the other hand, the workshops served to train the teachers addressing climate change impacts in class. Based on the feedback from teachers, an adapted version of the web portal was developed and finally launched in 2017. I participated in some of the workshops and other activities of the participative design process as a collaborator in the CIO project team. I also conducted interviews with the project team members, participated in the elaboration of guiding materials and co-edited an academic publication discussing the development process (Blumenthal et al. 2016).

As mentioned previously, the CIO platform turned out to be less self-explanatory than originally envisioned. The teachers also complained about a confusing setting of GUI elements:

> Figure 1 on the left also provides a first impression of the shortcomings of the graphical user interface design of the original portal. The small info box on the left-hand side has proved too small to truly aid orientation. The color legend on the bottom left-hand side can be easily overlooked. Also, the

40 Deutsche Bundesstiftung Umwelt (DBU). www.dbu.de, retrieved on May 7, 2019.

> original portal offered two separate helping systems (one can be activated by the bottom left-hand side ('i') and the other by the top ('?')). This has been a constant cause for confusion as users found it hard to find the information they required. (ibid.: 4)

The teachers in the workshops also criticized the high amount of technical and scientific terminology: "[...] a reduction of the scientific language was identified as a pressing issue" (ibid.). The project team invested a lot of work in solving these problems by redesigning and shuffling around GUI elements, reducing the number of technical terms and translating scientific terminology into everyday language:

> [...] That was so obvious: 'RCP 2.6,' 'RCP 8.5.' – no one can remember that or build a connection to it. That is why we translated it to 'strong climate protection' and 'weak climate protection.' That way, people can still make sense of it, even if they are not so familiar with the matter." (Interview with Irina Ballhaus)

The extent of the design changes possible, however, was limited by the structure of the underlying GIS. This can be illustrated with the discussion about a design element that did not make sense in the GUI, but couldn't be removed.

> Simon Hirsbrunner: I noticed that it is still there.
>
> Irina Ballhaus: Yeah, it's still there. In brackets. That's because of these presets [...]. We don't decide on that, it's what the computer is pulling out of it. It's automatic.
>
> SH: So you can't take it away?
>
> IB: You can't take it away. That's the way it is. Yes. [laughing]
>
> (Interview with Irina Ballhaus)

The platform generally turned out to be an enabler and a delimiter of all activities within the educational project. On the one hand, the original platform provided the starting point to think about further activities addressing the issue of local climate change impacts.

Figure 29: Comparison of old (left) and new (right) GUI of CIO.
Source: Blumenthal et al. (2016)

The technological availability of and experience gathered with the CIO platform had ultimately convinced funding agencies to support the pedagogic project, building on top of what already exists. On the other hand, this technological path dependency (Mahoney 2000) also narrowed the scope of the design process, with all the activities imagined to be strongly predetermined by the affordances of the existing platform. As several authors have shown, engineers often import solutions from one infrastructure to the next (Hughes 1983; Star/Bowker 2006: 232f). These solutions have been carried from WetterOnline's GIS into CIO, and from there into the educational project.

Acknowledging some of these concerns, the project team did not only engage in design improvements but also in tactics to work around the GUI. On the one hand, this entailed suggestions of navigation workarounds within the platform logic.

> Some of these suggestions could not be realized due to budget constraints, but it was possible to offer some workaround options: users can make screenshots or open an additional browser window for the cross-examination of individual maps. (Blumenthal et al. 2016: 4)

On the other hand, a number of additional materials were developed to mitigate the shortcomings of the online platform. This included a YouTube video explaining the CIO platform, a written guideline to teach climate change in class and sixteen units of teaching material focusing on particular thematic issues.

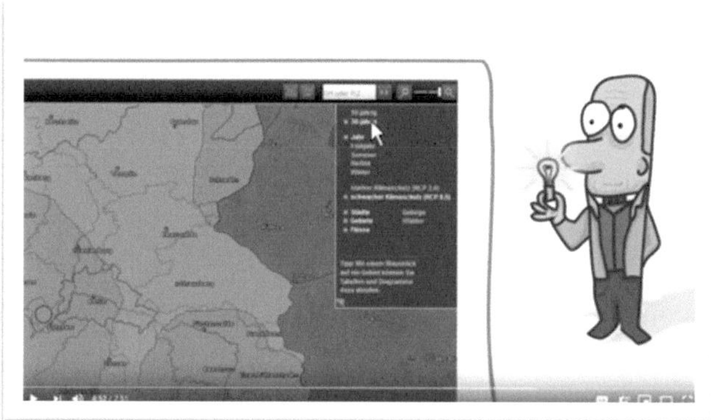

Figure 30: Screenshot from the CIO video tutorial on YouTube.
Source: PIK

The YouTube video was produced with a whiteboard video software (www.videoscribe.co) which lets producers drag and drop visual elements and arrange them within the timeline of an animation. The storyline entails a short introduction to climate change and a detailed walkthrough of the functionalities of the CIO platform:

> To provide an overview of the most important functions and how to navigate the portal, a tutorial (YouTube) was created using an animated introduction video. This tutorial is intended to help all user groups to easier access the functions and contents of the portal. (ibid.: 2)

The guidance paper (Blumenthal et al. 2016) addressed general issues regarding climate change communication; this includes the communication of uncertainties, connecting climate change to everyday life and facilitating behavioral change.

> Where can information on climate change be found? Which climate changes can be observed already today? How certain are projections about climate change? How can personal references to climate change be established? How can options for action be shown? (ibid.: 8)

Furthermore, comprehensive teaching materials were developed, which were tested and improved on over the course of the project period:

> In order to support teachers in using the portal, a variety of 16 teaching units were developed. The six "research workshops" were designed for individual sectors and are available in three different levels of difficulty. These levels may apply to different school or competence levels among students. Teachers can choose the one appropriate for their students' abilities. The "research workshops" can either be used during regular lessons or within interdisciplinary project work. Other teaching units are subject-specific. They have been designed for geography, but also for natural science subjects, mathematics or English lessons." (ibid.: 10)

In order to link to the community of teachers, the teaching units were uploaded to lehrer-online.de, an online platform providing a high number of teaching materials to educators. In its self-description, Lehrer-Online is

> the leading editorially supported material and service portal for teachers from all school types and levels. It focuses on tried and tested teaching units and materials that you can use in class without major preparation. In

> addition, Lehrer-Online offers you many innovative tools and functionalities that make your everyday life as a teacher easier. [41]

In the perspective of the CIO team, all of these artifacts were 'additional,' 'introducing,' 'giving an overview,' 'helping,' 'providing background information,' 'explaining' or 'supporting teachers in using.' By contrast, the geoplatform CIO has always been understood to be the main interface providing access to all relevant scientific knowledge. In the following, I would like to propose a different reading of the activities carried out within the PIKee project. This repositioning was enabled through a relational analysis limited to observable media practices during the temporal window of my fieldwork. Similar to the situational map of the Long Night of Sciences, I mapped the relationships between different elements (human, physical, symbolic) interacting within the PIKee project (see Fig. 31). The elements identified included pupils, teachers, scientists, science communicators, maps, diagrams, background texts in CIO, video tutorials, guidelines, teaching units, websites, TV news and street demonstrations. They also include institutions and infrastructures, such as the internet, the Potsdam Institute and the German school system. Accordingly, I have drawn lines for every interaction observed, mutually updating the set of elements. For the sake of readability, the situational map only depicts a selection of all the elements of analysis.

41 https://www.lehrer-online.de/ueber-uns/ last retrieved on May 29, 2019. Text translated by the author.

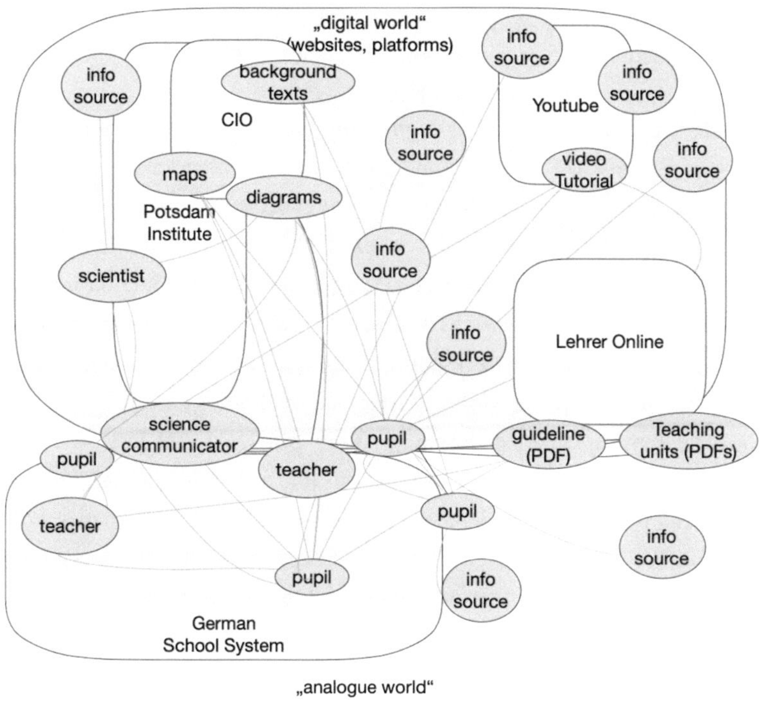

Figure 31: Interactional space of the educational project about CIO.
Source: My own visualization

The co-construction of technology and its users

We can characterize the design process of the educational PIKee project as a successful effort in building a sociotechnical infrastructure for the negotiation of climate change in schools. Susan Star and Geoffrey Bowker have shown that such *infrastructuring* entails a standardization of heterogeneous elements, including people, machines and symbolic artifacts:

> It is not just the bits and bytes that get hustled into standard form in order for the technical infrastructure to work. People's discursive and work

> practices get hustled into standard form as well. Working infrastructures standardize both people and machines. (2006: 235f)

While the CIO had always been presented as a working tool for 'users,' it actually took a lot of time and work to create a community with an especial practice and routine to 'use' the platform together with many other artifacts and techniques of knowledge acquisition. Nelly Oudshoorn and Trevor Pinch (2005) wrote in the introduction to their anthology *How Users Matter: The Co-construction of Users and Technology* that the design process of digital technologies always entails a co-construction and reconfiguration of users and technology.

The key aspect in co-creating 'users' was to focus less on information retrieval from the CIO platform and more on the daily media practices of teachers and pupils in the setting of the German school system. We can exemplify this with the routine of a teacher preparing and performing two hours of teaching addressing the topic of climate change.

> Teacher Nr. 2: You have to see it like this: It's Sunday afternoon. I'm sitting there and want to finish this quickly. [...] Well, we don't spend hours looking at it. That must be fast and self-explanatory. (Teacher testimony in a CIO workshop at a School in Berlin, translated by the author)

The preparation and teaching as a sequence of interactions may be formalized as the following sociotechnical script:

Day 1 (preparation):

- The teacher browses through Lehrer Online to explore possible topics to be addressed in class.
- The teacher decides to address climate change in the class and enters 'climate change' or 'environment' into Lehrer Online's search console.
- The teacher finds PIKee and CIO as a result of the query.

- The teacher explores CIO, watches the YouTube tutorial, reads the guidelines and makes him/herself familiar with the worksheets.
- The teacher adapts the worksheets to his/her teaching subjects and course structure.

Day 2 (teaching):

- The teacher introduces the topic of climate change in class.
- The teacher gives pupils time to explore the material available, including CIO, the YouTube video and the worksheets.
- Pupils try to solve the problems given in the worksheets.
- The teacher and pupils engage in an open debate about the future with climate change.

The realization of this routine took four hours, including the teacher preparation and actual teaching experience. Acknowledging this time window (four hours), we can make an assessment of the time actually spent interacting with people and different kinds of artifacts. The actual 'screen time' on CIO appeared to be rather short. The time was equally distributed between browsing the platform, reading through background material, watching the YouTube video, filling out the thematic worksheets and debating aspects of climate change broadly.

An Anchoring device

Does this mean that the CIO platform could have been exchanged with another easy-to-produce video or a simple webpage collecting information resources on climate change? This is certainly not the case. The features of the platform were key to enabling the ultimate goal of the educational project PIKee – the facilitation of knowledge acquisition and debates about climate change impacts. I would argue that this role of the platform could be characterized as an *anchoring device*. This role bears some similarities with the description of Henderson's conscription device, which has been discussed previously. Building on Lucy

Suchman's (1988) description of the whiteboard's role in organizing work processes and Susan Star and James Griesemer's boundary objects (1989), Henderson refers to engineering sketches as 'conscription devices'[42] as they

> enlist the participation of those who would employ them in either the design or production process, since users must engage in the generation, editing, and correction of drawings during their construction if the design is to serve its intended function. (Henderson 1991: 452)

In a similar way as the sketches enable discussions of engineering processes in addition to WGs, the CIO facilitated debates about climate change among heterogeneous actors. The CIO enlists the participation of these actors for a debate of climate change impacts in an educational setting. In contrast to the conscription device, however, the CIO platform was an element that ran in the background of the interactions between participants of the education activities. Rather than enlisting people, it anchored and channeled the activities around climate change impacts. I could witness in the workshops and other activities within the project that debates often got lost in the complexity and versatility of the issues at stake. While such excursions into the unknown were fruitful to open debate, they threatened the frame of the educational setting and attributed roles of the participants. The teachers were all committed to address climate change in their curriculum, but they were haunted by the controversies around climate change and the fear of losing control in their classroom. The CIO platform was a way to anchor the discussions around 'the current state of science,' represented by a discrete container of maps, diagrams and accompanying text. Accordingly, it was unproblematic if debates drifted away from the scientific facts as they could always be navigated back to the platform. Consequently, it was possible to engage in open debates regarding the pupils'

42 Taken from an understanding of 'military conscription.'

everyday life perceptions and experiences of climate change. In other words, participants could transcend the boundaries of established practice in the class room without perceived border transgression.

A packaged body of knowledge

The role of the CIO as an anchoring device in the educational activities of PIKee was enabled through a number of characteristics that build trust among actors. What became clear during my analysis is that these features appear paradoxical once unraveled. However, this paradoxical nature is exactly their source of strength. The fluid symbolic outputs of such media technologies can mean different things to different people. This interpretative flexibility enables them to mobilize actors with very different attitudes towards climate, science and technology.

First of all, the homogenous aesthetics in the platform mediates a stability of the scientific facts contained. As we have seen previously, the harmonization of the underlying heterogeneous datasets for the depiction in the GIS has been cumbersome work, which is black boxed by the singularity of the CIO interface. Susan Star says that the CIO enables the construction of "a single voice that does not problematize diversity," which "speaks unconsciously from the presumed center of things" (Star 1999: 384). The construction of such a single voice would not be possible through the assembly of diagrams, maps and texts from exterior resources. It is exactly the singularity of the aesthetical feel and navigational logic that confines harmony. This single voice is not disturbed even by the fact that the underlying data has been repeatedly updated to reflect new scientific insights and improvements to former simulation outputs.

Secondly, the interactivity of the interface creates a potentiality for checking all the data by hand. A suspicious person might zoom through the maps, slide through the timeline and point at different scenarios without identifying any irregularities. The data seems complete, and there are no apparent blind spots on the maps. The proximity of the navigational logic and aesthetics of the interface to those of simulation modeling practice creates a perceived circulating reference between the facts depicted and the underlying scientific processes. In other words, data in CIO 'talks on its own.' We will see later (Chapter V) that this talking ability of 'raw data' is a fallacious imaginary. At this stage, it is only important to highlight that it must be characterized as a potentiality and a promise. While no one will actually check all the data in the platform, it suffices that the possibility is given by technological design. It was crucial in the activities observed within the PIKee project that the data could potentially be 'realized' within seconds in the concrete situation, namely, as a visible and understandable choropleth[43] map. To use a (potentially problematic) analogy: A person visibly carrying a gun necessarily transforms the relationships in a situation. This is independent of the weapon eventually being fired or not.

Thirdly, the fact that the entire body of knowledge in the CIO container comes from one institution established a well-defined actor of trust for the knowledge packaged. Here, we may understand trust as a tactic for the reduction of social complexity, as argued by Niklas Luhmann (2014). The individual scientific facts in the container are black boxes, which cannot be realistically situated or checked by outsiders of the simulation processes. The single producers of the scientific facts cannot be identified easily, which leaves the institution of the PIK, and climate science more generally, as the only actors to be trusted and challenged.

43 Wikipedia entry: "A choropleth map (from Greek χῶρος 'area/region' and πλῆθος 'multitude') is a thematic map in which areas are shaded or patterned in proportion to the measurement of the statistical variable being displayed on the map, such as population density or per-capita income." Retrieved on April 23, 2019, via https://en.wikipedia.org/wiki/Choropleth_map.

IV. Future models

The field site of this chapter lies in the realm of the digital and deals with technological entanglements in regional impact modeling. Regional climate modeling is an intriguing scientific practice, because it seems to reverse climate science's obsession with the global scale. As Paul Edwards has shown, meteorologists have fought for centuries to "make global data" (building a global observation and communication infrastructure) and to "make data global" (standardize heterogeneous datasets) (2010). This infrastructural work was a major achievement and enabled the discovery and scientific proof of climate change. What are the reasons for the renewed interest in making global data local again or, as Mahony (2017) puts it, "the (re)emergence of regional climate"?

Global-to-local

From the perspective of simulation modelers, regionalization is mainly a matter of resolution. We can illustrate both these concerns by a description of RCMs on the website of the CORDEX[44] project, a globally coordinated effort in downscaling global models to regional scales:

> Global Climate Models (GCM) can provide us with projections of how the climate of the earth may change in the future. These results are the main motivation for the international community to take decisions on climate

44 Coordinated Regional Climate Downscaling Experiment.

change mitigation. However, the impacts of a changing climate, and the adaptation strategies required to deal with them, will occur on more regional and national scales. This is where Regional Climate Downscaling (RCD) has an important role to play by providing projections with much greater detail and more accurate representation of localised extreme events.[45]

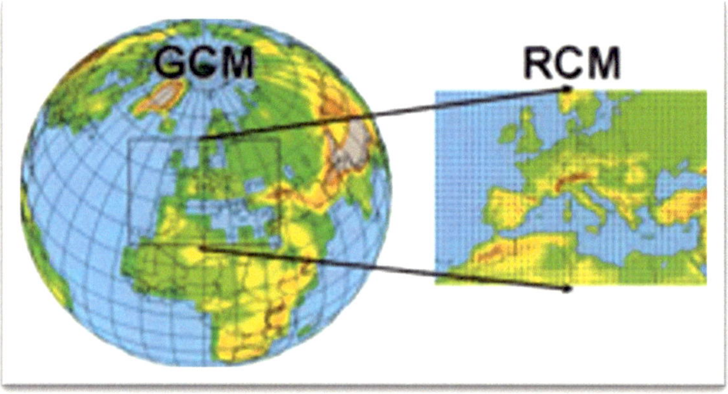

Figure 32: An RCM domain embedded in a GCM grid.
Image source: Giorgi (2008)

There are three established methods to create local climate data: (1) Increasing the resolution of a global model, (2) running a statistical regional model, which derives local data from the output of a global model, or (3) setting up a dynamic local model with its own physical logic (Mahony 2017: 140ff). All of these methods have their strengths and weaknesses, which then also define possible application fields. Regarding their representational logic, the simulation of regional climate is different from global modeling, in the sense that regional models must represent phenomena, such as high- and low-pressure areas, vegetation, land use, glaciers and snow cover. And, as climate scientists often argue, regional models must be able to simulate in a higher

45 https://www.cordex.org/about/what-is-regional-downscaling/, retrieved on May 5, 2019.

resolution than global models in order to represent these processes accurately. Meteorological models of weather prediction need to perform similar tasks and enable verification regarding empirical observations; regional modeling builds mainly on model structures in meteorology (Jacob et al. 2017: 29). We may briefly compare two of these methods – statistical downscaling (2) and dynamic downscaling (3) – to illustrate some of these differences.

Dynamic downscaling

Models of dynamic regionalization calculate climate impacts with a three-dimensional excerpt of the atmosphere; this is similar to the global models but with a higher resolution. Dynamic models basically resolve a theoretical system of equations on a defined spatiotemporal grid. The equations represent laws of conservation for energy, impulse and the mass of the air, as well as water and water vapor. The dynamic regional model starts with the outputs of a global model and obtains new boundary values from the latter every six hours (in simulated time). Consequently, the global model is also formative for the long-term variability and the large-scale processes of the region of the model (e.g. Europe). The regional climate is then calculated gradually by increasing the spatial resolution of the model: Firstly, to a grid mesh width of 50 km and then down to 10, 7, 3 or even 1 km. The higher spatial resolution enables the representation of characteristics of the Earth's surface, such as the altitude structure and land cover, and processes, such as local precipitation and cloud coverage. Dynamic modeling is often described as the "royal road" (Orlowski 2007: 3).

Statistical downscaling

A statistical regional model works differently. Statistical modeling explores relationships between large-scale weather conditions or global circulation patterns and local climate data. The statistical model STARS, for example, developed at the Potsdam Institute, rearranges the time series of climate variables observed and simulated in order to take into account prescribed, linear trends. The results are synthetic, comparable time series of meteorological variables at the places of meteorological weather stations. As a result, the resolution is determined by the spatial density of weather stations. The advantage of statistical models is that they need a lot less computing time than the dynamic models. However, they are unable to simulate events that are fundamentally different or more extreme than those observed in the past because they are literally mirrors and conditioned projections of the past into the future (translated and summarized from Jacob et al. 2017: 28f). The PIK's statistical model STARS has triggered a number of scientific controversies due to these representational flaws. The model has been operationalized to simulate future climate change in Germany, making statements about developments at the spatial scale of counties and districts. Further down the model chain, STARS has been used to drive a number of climate impact models which simulate the consequences of precipitation and water systems (floods, extreme weather), agriculture (drought, flowering times, cultivation of new wine grapes), tourism and health-related issues (heatdays). The simulation outputs triggered a variety of discussions in the mass media[46] and provided a scientific base for interdisciplinary studies making sense of the future with climate change in Germany (Gerstengarbe et al. 2013). Ten years later, scientists at the PIK challenged the mathematical logic of the STARS model and some interpretations of the simulation outputs (Wechsung/Wechsung 2016, 2015). Inter alia, the new reconsiderations have been possible thanks to experiments migrating STARS to other regions, such as the Chinese Guanting region (Wechsung/Schellnhuber 2018). Myanna Lahsen showed that modelers tend to be very protective of their own models, given the long time they spend 'raising' them (2005).

46 See, for example, https://www.welt.de/wissenschaft/umwelt/article5456480/Wie-der-Klimawandel-Deutschland-trifft.html.

However, the episode concerning STARS also exemplifies the sophisticated self-correcting mechanisms in place within the climate sciences.

The main strategy of climate scientists to account for the uncertainties of different models and their procedures is to compare and average them within model ensembles, similar to that in global climate modeling practice. Coordinated efforts, such as CORDEX, then produce coordinated sets of regional downscaled projections for all the regions of the world. As a result of the ensemble process, scientists engaging in further modeling at the local level (e.g. climate change impact modeling) do not have to bother with choosing between different models or their outputs but can rely on standardized climate time series data that drive their own predictions of the future. However, this also means that impact modelers have to trust the soundness of the original downscaling models, simulation process, averaging methods and data output mechanisms. Paul Edwards has shown that the *vast machine* of climate science is the extraordinary instance of trusted infrastructure.

Place-to-place

As Martin Mahony has shown, regional modeling can be described as a practice of translation. For him, RCMs broadly fulfill two functions: On the one hand, they are employed to "re-invest the global climate with some of the local meaning of which it is stripped in the moment of its construction" (Mahony 2017: 140). On the other hand, RCMs are key tools for nation-states translating climate change into something they can govern:

> National maps of climate impacts re-territorialize climate change, and enable states to perform a competent engagement with risks and uncertainties that are paradoxically beyond their own capacities of control. (ibid.)

Once a regional model is technically developed, the question arises to what extent it is location-specific or if it can be easily operationalized for multiple geographic spaces. In other words, the question is how much the map (model) resembles a territory. The human geographers Mike Hulme and Martin Mahony have investigated such questions of mobility and mutability of local climate models in a number of articles (Hulme 2008; Mahony 2017; Mahony/Hulme 2012). As Mike Hulme has put it, climates do not seem to travel well between scales:

> It is important to notice what happens in this circuit of transportation. Weather is first captured locally and quantified, then transported and aggregated into regional and global indicators. These indicators are abstracted and simulated in models before being delivered back to their starting places (locales) in new predictive and sterilised forms. 'Digitised' weather for virtual places can even be conjured from these models using stochastic weather generators. Through this circuitry, weather – and its collective noun climate – becomes detached from its original human and cultural setting. (2008: 7)

It appears convenient to investigate issues of scale and mobility in the context of regional climate modeling, considering that such geographic mobilization is often an explicit goal of these research endeavors. Can the German model be migrated to China, the Elbe model to the Yangtze river, as a movement from one represented Euclidian space to another? As a side note, it may be added that there have also been attempts to apply global climate Earth models to other planets in order to learn about their climates and about ours (Kasting et al. 1988) The prime case is the investigation of the Venus syndrome (Goldblatt/Watson 2012), as a model for 'climates gone bad' or 'runaway climate change.' Nevertheless, the most obvious practical case for climate-model migration is from one geographic place to another.

Mahony and Hulme aim at assessing, "how scientific tools are able to overcome the friction of distance and attain 'usefulness' in new places, and the effects of these transfers on the epistemic landscapes of their new environments" in their investigation of the PRECIS (Producing Regional Climates for Climate Impacts Studies) model (2012: 198). The latter is an RCM developed by the United Kingdom's Met Office Hadley Centre. According to the Hadley Centre's website, it is

> [...] a regional climate model (RCM) that takes large scale atmospheric and ocean conditions from observations or global climate models (GCM) where horizontal resolutions vary from 100 to 300 km, and downscales it over a region of interest to resolutions of 25 or 50km. This allows for a more realistic representation of the climate over the region of interest, accounting for complex surface features such as mountains, coastlines and islands which are not resolved in the global models. [47]

The PRECIS is an accustomed traveler. While originally conceptualized for European territory, the system has migrated to a variety of places, including India and South Africa. It has not only toured throughout the world but has also been in the hands of a variety of different actors. Building on Anselm Strauss' (1978) social worlds concept, Mahony and Hulme argue that PRECIS facilitates interaction and exchange between multiple worlds and sub-worlds:

> [...] PRECIS can be seen to facilitate interaction and exchange between a number of worlds and sub-worlds: model developers, the climate impacts community, global and national political assemblages, non-governmental institutions etc. The climate arena, within which the various actors interact, provides a transaction space [...] whereby asymmetrical relationships of dependency can develop at institutional and disciplinary boundaries. (Mahony/Hulme 2012: 208)

---

47 https://www.metoffice.gov.uk/research/applied/international/precis/introduction.

The authors also characterize PRECIS as a *boundary object,* building on Susan Star and James Griesemer's (1989) conceptualization:

> PRECIS' multivalent purposes, as articulated by developers, partners and users, its notional flexibility engendered by its mobility, and its ability to fulfill a range of substantive, instrumental and discursive demands make it eligible for this description. (ibid.: 208)

As multifarious as RCMs may individually behave as politico-scientific devices, they are gradually becoming *obligatory passage points* (Callon 1984) for the accomplishment of a political sagacity (Mahony/Hulme 2012 208).

> This is achieved through the translation of instrumental goals and the deployment of normative discourses of vulnerability and scientific realism, the consequence being a community pursuing knowledge that possesses high spatial resolution and precision. This pursuit is facilitated by the rendering of planned adaptation as captive to, or an ancillary of, the ability to predict future climatic changes on the scales that most interest decisionmakers. (ibid.)

Mahony and Hulme highlight a number of consequences of the establishment of RCMs as obligatory passage points. Notably, the recurrence to (a certain type of) models may privilege some approaches of climate adaption strategies over others (optimal, rather than robust) (Dessai/Hulme 2007). However, Mahony and Hulme also draw more general conclusions from such preferential treatment of model predictions over other practices to think about the future. For these authors, they represent an "unfolding geography of epistemic power" with climate as a "chief determinant of humanity's putative social futures" (Hulme 2011; Mahony/Hulme 2012).

Such climate determinism and reductionism are also considered and debated in the community of climate impact modelers at the PIK. Many impact modelers are not trained as climatologists but as economists, agriculture specialists and hydrologists. As has been mentioned previously, they have to rely on climatologists to drive their own projections of the future:

> TC: Yes, so the problem is, we usually only see the future from a climate perspective. There are very good climate models now, compared to the ones we had a few years ago. [...] So, we have a variety of scenarios that we can feed into it as different realizations of future climate. And we just look at how the [water] runoff behaves. However, change in land use is much more difficult to describe. So we haven't done it for this area. (Interview with Torsten Casius, translated by the author)

In this sense, climate models do not only project future climates but also imprint a view of other future developments, such as land use changes and urbanization. Put positively, climate change is also a mobilizing element for other environmental knowledge, then traveling into the future as a free rider. Climate science provides a spatiotemporal grid for the future, which can then be colored by impact modelers and other actors. It might, therefore, be understandable that impact modelers often maintain a controversial relationship with climate models and modelers.

Infrastructural migration

The existing literature in STS has treated regional and impact modeling as a matter of knowledge translation raising representational issues. This equally includes considerations of epistemological and political representation.

> [...] scientists, campaigners, and politicians have long been aware of the politically paralyzing effects of knowledge claims that refer to abstract, global realities rather than the local realities of everyday existence or routine political decision making. (Mahony 2017: 139)

By contrast, model migration can also be understood as a technological issue, migrating a model from one machine, system or infrastructure to another. In one instance, Martin Mahony briefly discusses the computational practices and infrastructures making models travel. Common issues of epistemic uncertainty and model opacity in regional modeling are usually countered by choosing open-source software tools. However, Mahony claims that the trend toward open-source simulation software and models, accompanied by a rhetoric of transparency and reflexivity, is, in fact, characterized by a high degree of epistemic opacity. According to him, simulation modelers often do not truly understand the design principles and assumptions of the simulations and models at stake. This *epistemic opacity* of modeling technology (Kouw 2010: 4) is strongly dependent on factors such as the detail of the accompanying handbooks and the user's trust in the scientific credibility of the model constructors (Mahony 2017: 152). Drawing on Matthijs Kouw's extended interpretation of the term 'vulnerability' [48] (2010: 1) and his analysis of modeling technology in hydrology, Mahony argues that epistemic opacity can create a kind of vulnerability based on software design and use:

> In the case of PRECIS, this "epistemic opacity" was a product both of the desire to produce a usable tool, and of the wish to preserve the authority of the Hadley Centre's own development and coding. PRECIS travels the world through a network of national contact points who receive training from the

48 Vulnerability is a key term in climate impact research. While many contesting definitions of the concept exist (Füssel 2005), the IPCC characterizes vulnerability as follows:

"The degree to which a system is susceptible to, or unable to cope with, adverse effects of climate change, including climate variability and extremes. Vulnerability is a function of the character, magnitude, and rate of climate variation to which a system is exposed, its sensitivity, and its adaptive capacity." (McCarthy/IPCC 2001: 6).

> Hadley Centre with support from governmental agencies. For one scientist not associated with the Indian contact point in Pune, PRECIS has been an inaccessible tool despite arguments that the model should be run at more than one location in India [...]. (Mahony 2017: 152)

Epistemic opacity in regional climate modeling for Mahony is then shaped by several factors, including software interfaces that hide the model's core code, expert considerations and restrictions regarding code access, and the material realities of scientists in resource-poor (and, thus, computationally limited) institutions and/or countries (ibid.:155). In this reading, the discourse of open software promises the mobilization of computer models, their stored knowledge and instrumental capacities to various geographic places, user communities and technical systems. However, these promises are only partially met for Mahony and fail to mitigate the epistemic opacity of computer models and modeling as a techno-scientific practice.

Against this view, I would argue that scientists indeed manage to address epistemic opacity quite successfully with a number of techniques that are discussed as follows.

Investigating a model

This second part of the chapter will address technological practices of mobilization in climate impact modeling, which I will refer to as *mobile modeling*. I will discuss these recent practices in impact modeling using the empirical example of CLIMADA (CLIMate ADAptation), a climate risk assessment and damage calculation model, tool and platform. The software stack and infrastructural entanglement around CLIMADA can be exemplary for shifting practices in climate impact modeling and scientific software development in general. I came across CLIMADA while interviewing Tobias Geiger, a PIK expert for the modeling of extreme weather events, hurricanes and particularly their economic damages. Tom is not the main developer of CLIMADA but a contributor to its code, where he added a module for the simulation of hurricanes and their damage caused to local economies. The model itself (CLIMADA) has been developed and maintained by David Bresch, Professor for Weather and Climate Risks at the Swiss Federal Institute of Technology in Zurich (ETHZ). The user manual of the model states that "CLIMADA is an open-source and -access global probabilistic risk modelling and adaptation economics platform." (CLIMADA Manual: 1) It aims at strengthening weather and climate-resilient development and providing decision-makers "with a fact base to understand the impact of weather and climate on their economies, including cost/benefit perspectives on specific risk reduction measures" (ibid.). The functionalities of CLIMADA can be illustrated by two visualizations from that part of the model. Based on spatiotemporal data representing the distribution of 'assets' (e.g. buildings, agricultural areas) at a geographic location, CLIMADA simulates the economic damage of natural disasters to these entities. The map in Figure 34 depicts the risk exposure of assets as green spots (existing but low exposure) and a few red ones (high exposure). One can specify different types of disasters (e.g. a hurricane or a Tsunami), time frames or territories by manipulating input data and functions of the model.

CLIMADA can represent historic events or simulate future ones, the latter based on projections of socioeconomic and climate variables. It also makes a prediction about the share of additional damage caused by climate change, with Figure 34 showing a possible result of these calculations on the right. It prognosticates an accumulated risk of damage of about 34 billion USD until the year 2040. It also anticipates that a large faction of this damage (21 billion USD) will be attributable to climate change.

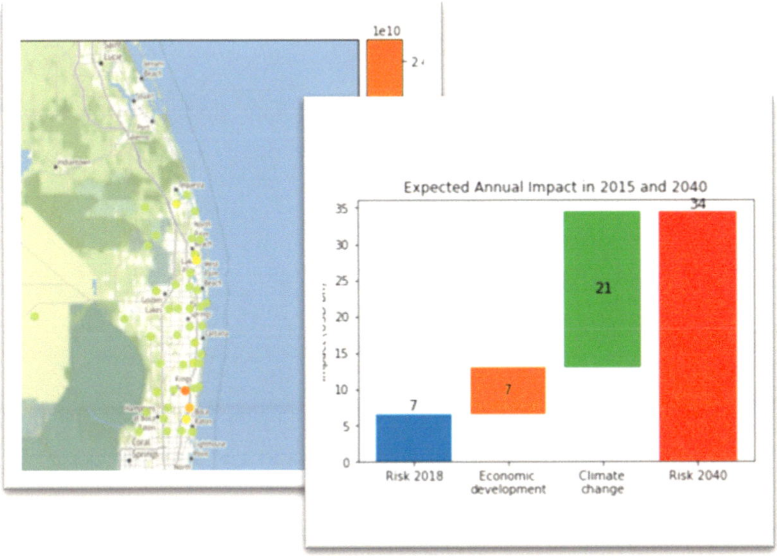

Figure 33: Mapping exposures to natural hazard at a location within the US Federal State of Florida. Source: CLIMADA Jupyter Notebook

Access log

I developed the methodological device of an *access log* to trace the infrastructural elements of CLIMADA. The term is inspired by the use of the word in ship navigation and software development. A logbook or log in shipping is a record of important events in the management, operation and navigation of a ship[49] Software development draws from this use of the word in shipping and translates it to the situation on the web:

> A Web log file records activity information when a Web user submits a request to a Web Server. The main source of raw data is the web access log which we shall refer to as log file. As log files are originally meant for

49 https://en.wikipedia.org/wiki/Logbook, retrieved on April 3, 2019.

> debugging purposes.
> (Suneetha/Krishnamoorthi 2009: 327f)

In my case, the access log is a special 'field-note' that traces the steps to set up the model on a local machine (laptop).

- open github.com website
- type 'CLIMADA' into the search field
- open the project page
- tap 'clone or download'
- wait 15 seconds (file is downloading from github.com to my laptop)
- click on the zip-file 'climada_python-master.zip' in the downloads folder of my laptop (file unzips)
- copy CLIMADA folder into Python folder structure on my laptop
- open Anaconda Navigator (click on application alias) on laptop and launch the anaconda IDE (integrated development environment)
- Launch Python Jupyter notebook
- Terminal opens and launches Python on laptop
- Open readme file in notebook
- Read installation instructions
- Open linked guide[50] for more info
- As indicated in guide, install dependencies in Anaconda by choosing Environments/Import
- Anaconda creates a new software environment for CLIMADA (takes 5 min)
- In Jupyter notebook, navigate to climada_python-x.y.z. repository and open doc/tutorial/1_main_climada.ipynb file
- Jupyter launches CLIMADA notebook in browser (Firefox)

50 https://climada-python.readthedocs.io/en/stable/guide/install.html, retrieved on May 6, 2019.

The access log helped me to identify infrastructural elements and relationships within and toward technologies of interest. It helped, for example, to identify Github, Python, Anaconda, specific libraries and Jupyter Notebook as entities to be considered in my research. One could characterize these elements as 'dependencies'[51] of a specific technology or infrastructure.

This entire process of setting up CLIMADA takes about 30 min on an Apple Macbook Pro (2017 model) laptop. It could equally be installed on a Microsoft or Linux machine taking the same steps. Of course, a successful setup also comes with sociotechnical preconditions. One had to learn Python, getting to know Jupyter Notebooks, installing Anaconda, Python, and its libraries on the local machine. One needs a functioning internet connection. It is useful to know where to find things on Github and to tab communities at Stack Overflow for troubleshooting. However, it is still impressive that formerly highly esoteric technologies, such as climate models, are, or appear at least, relatively open and accessible with broadly disseminated (programming) skills. In the following, I will focus on some of the elements and relationships identified in the access log, as they particularly characterize contemporary scientific programming and *mobile modeling* especially.

Coding openness

The CLIMADA was originally developed in MATLAB,[52] a proprietary program owned by the US-American company *Mathworks,* which specializes in mathematical computing software. However, in 2017, it was

51 This use of the word differs from the one in computer science. Here, dependencies are literally the external pieces of code that have to be called by a specific program.

52 The official website of MATLAB, retrieved on June 4, 2019, via https://www.mathworks.com/products/matlab.html.

decided to translate the whole CLIMADA code into the Python programming language and to make it available as open-source and free software. The exiting MATLAB version, by contrast, will no longer be maintained.[53] 'Open-source' and/or 'free' means that software source code is equipped with a specific legal license, in this case, a GNU lgpl (Lesser General Public License).[54] It is beyond the scope of this study to discuss the different versions of open software licenses, even if this licensing choice has a strong impact on the meaning of 'openness' and the politics of amplification in question. In a historical perspective, it must be said that opening programming code is not a new practice in software development but goes back to the origins of the craft in the 1960s and 1970s, as described by von Hippel and von Krogh:

> In the early days of computer programming commercial "packaged" software was a rarity – if you wanted a particular program for a particular purpose you typically wrote the code yourself or hired it done. Much of the software development in the 1960's and 1970's was carried out in academic and corporate laboratories by scientists and engineers. These individuals found it a normal part of their research culture to freely give and exchange software they had written, to modify and build upon each other's software both individually and collaboratively, and to freely give out their modifications in turn. This communal behavior became a central feature of "hacker culture." (2003: 3f)

53 Information from CLIMADA Github page, retrieved on June 6, 2019, via https://github.com/davidnbresch/climada.

54 Wikipedia page of the GNU LPGL license agreement:

"The license allows developers and companies to use and integrate a software component released under the LGPL into their own (even proprietary) software without being required by the terms of a strong copyleft license to release the source code of their own components. However, any developer who modifies an LGPL-covered component is required to make their modified version available under the same LGPL license." Retrieved on June 4, 2019, via https://en.wikipedia.org/wiki/GNU_Lesser_General_Public_License.

Such practices have come with a promise that everyone can potentially use the software, which is also true for the CLIMADA collaboration between ETHZ and PIK researchers:

> It's just that the whole basic structure exists in MATLAB. Which, in my opinion, is not very user-friendly. Because, on the one hand, there are license fees. And, on the other hand, it is not so user-friendly for me. Mmm. But yes, that will perhaps also change in the future. So, there are already the ideas that maybe you can import this complete package into Python and then use it there. (Interview Geiger, translated by the author)

The interview with Geiger was carried out at the beginning of the cooperation with the ETHZ. Two years later, the whole code had been translated from MATLAB to Python. In this sense, the practices of 'mobilization' of CLIMADA had been successful. Initially, opening the source code triggered an engagement of the PIK researchers to contribute to the ETHZ software. On the other hand, embedding the PIK module in a broader modeling endeavor (CLIMADA) also amplified the impact of the work carried out in Potsdam.

As this example shows, 'coding openness'[55] is more than a license issue by far. It includes a variety of ideas, practices and infrastructures, some of which are discussed further below.

Pythonization

The shift from proprietary to open-source programming languages and environments translates particularly to a transformation we might refer to as *Pythonization.* Scientific models are increasingly imagined and formulated in Python, a high-level language such as C, C++, Perl and

55 Term taken from Prof. Dr. Claudia Müller-Birn's course taught at the Computer Science department of Freie Universität Berlin. More info on https://www.mi.fu-berlin.de/en/inf/groups/hcc/teaching/summer_term_2019/coding-openness.html, retrieved on June 5, 2019.

Java. As computers can only directly read low-level languages ('machine' or 'assembly languages'), programs written in a high-level language have to be processed before they can run. This extra processing takes some time, which is a disadvantage of high-level languages. However, the latter also carry enormous advantages. *Think Python: How to Think Like a Computer Scientist*, a popular textbook introducing the Python programming language to prospective users, puts it as follows.

> First, it is much easier to program in a high-level language. Programs written in a high-level language take less time to write, they are shorter and easier to read, and they are more likely to be correct. Second, high-level languages are portable, meaning that they can run on different kinds of computers with few or no modifications. Low-level programs can run on only one kind of computer and have to be rewritten to run on another. (Downey 2012: 1)

We will come back to this feature of 'portability' in Chapter V. While this explanation clearly shows the advantage of high-level programming languages in comparison to assembly languages, it does not explain why Python is currently thwarting formerly dominant languages, such as Java or C++. We can find some arguments for its success in another Python reference textbook, *A Whirlwind Tour of Python* published in 2016. According to its author, Jake VanderPlas, "the appeal of Python is in its simplicity and beauty, as well as the convenience of the large ecosystem of domain-specific tools that have been built on top of it" (2016: xii). As for this "simplicity" and "beauty," we can find a spiritual self-description of these qualities built right into the heart of the language code by typing the command "import this" into a Python console:

```
import this

The Zen of Python, by Tim Peters

Beautiful is better than ugly.
Explicit is better than implicit.
Simple is better than complex.
Complex is better than complicated.
Flat is better than nested.
Sparse is better than dense.
Readability counts.
Special cases aren't special enough to break the rules.
Although practicality beats purity.
Errors should never pass silently.
Unless explicitly silenced.
In the face of ambiguity, refuse the temptation to guess.
There should be one-- and preferably only one --obvious way to do it.
Although that way may not be obvious at first unless you're Dutch.
Now is better than never.
Although never is often better than *right* now.
If the implementation is hard to explain, it's a bad idea.
If the implementation is easy to explain, it may be a good idea.
Namespaces are one honking great idea -- let's do more of those!
```

Figure 34: The Zen of Python. Source: My own screenshot, Zen by Tim Peters

On a more concrete and technical level, one thing that distinguishes Python from other programming languages is that it is interpreted rather than compiled: "[T]his means that it is executed line by line, which allows programming to be interactive in a way that is not directly possible with compiled languages like Fortran, C, or Java" (ibid.: 5). Such interactive coding enables extensions, such as the *Jupyter Notebook,* which revolutionize the way scientific programming works (see more below).

The trend toward Python in coding practice was a recurring theme in my interviews and discussions with impact researchers. Referring to Gabriele Gramelsberger's characterization of FORTRAN as the *lingua franca* of climate modeling (2008a: 144), we may argue that Python is increasingly taking up this position in climate impact research. The reasons for this shift go beyond the mere consideration of Python's qualities as a programming language. They are linked more to an infrastructural entanglement that includes elements such as multipurpose software packages and powerful community platforms.

Packaging code

The large ecosystem of domain-specific tools for scientific computing and data science is built around a group of modules[56] and packages[57] (also referred to as 'libraries'[58] in everyday speech). Common Python packages are NumPy (storage and computation for multidimensional data arrays), SciPy (numerical tools, such as numerical integration and interpolation), Pandas (a set of methods to manipulate, filter, group and transform data), Matplotlib (an interface for the creation of publication-quality plots and figures), Scikit-Learn (a toolkit for machine learning) and IPython/Jupyter (for the creation of interactive, executable documents) (VanderPlas 2016: 1f). The packages can easily be imported into one's own programming code, where they perform a variety of tasks for the structuring, analysis and representation of data. As a result, Python has become particularly effective for coping with the contemporary challenges of data-intensive science, 'Big Data' or 'data science':

> As an astronomer focused on building and promoting the free open tools for data-intensive science, I've found Python to be a near-perfect fit for the types

---

56 "Python has a way to put definitions in a file and use them in a script or in an interactive instance of the interpreter. Such a file is called a module; definitions from a module can be imported into other modules or into the main module (the collection of variables that you have access to in a script executed at the top level and in calculator mode)." From the Python documentation, retrieved via https://docs.python.org/3/tutorial/modules.html on July 6, 2019.

57 A package is a collection of modules. See Python documentation, retrieved via https://docs.python.org/3/tutorial/modules.html#packages on July 6, 2019.

58 In contrast to Java Script and other languages, Python does not formally entail 'libraries.' However, it is often used synonymously for modules and packages.

> of problems I face day to day, whether it's extracting meaning from large astronomical datasets, scraping and munging data sources from the Web, or automating day-to-day research tasks. (ibid.)

This aspect of automation also becomes increasingly important within climate impact research, as the following excerpt from an interview with a PIK scientist shows:

> So, the data comes from two sources, one HTML website (ratification) and a CSV (emissions). I wrote a Python script to extract the data. The export from the HTML page is automatic." (Interview Gatow)

While the majority of the work is still arranged around simulation models, impact researchers also increasingly engage in capturing live data, be it from the web or from updated servers providing satellite imagery. These practices pertain to what is now considered as *data-science* rather than computational science. In a loose understanding, the term data-science is often associated with the contemporary challenges of 'big data' and opportunities in machine learning, deep learning and artificial intelligence to deal with it. Conceptually, data science may grasp a more generalized shift in academia and industry to take 'data' as the primary object to be dealt with (Ribes 2018: 2), translating to an entanglement of data collection, engineering, analytics and representation (Computing Research Association 2016). Experts in the field have also highlighted the transdisciplinary nature of data science for academia, arguing that "across academic disciplines, the computational and deep data problems have major commonalities. If researchers across departments join forces, they can solve multiple real-world problems from different domains" (O'Neil/Schutt 2013: 15). A potential has also been seen particularly in 'AI for good,' with data science addressing specially socio-environmental challenges, such as climate change, biodiversity loss and natural risk prevention and management (Bundesregierung 2018: 17; International Telecommunication Union 2018:

26; Karpatne et al. 2017). Nevertheless, a variety of actors have also highlighted the potential risks of data science practices, linked to issues such as bias in and opacity of such algorithmic systems (AI Now 2018; Crawford 2013; Crawford/Calo 2016).

In any case, the aptitude for data-science explains some of the attractiveness of Python for actors outside the world of science:

> Conceived in the late 1980s as a teaching and scripting language, Python has since become an essential tool for many programmers, engineers, researchers, and data scientists across academia and industry. (VanderPlas 2016: 1)

Industrial actors building their products with Python include the big players of the data and platform economy, such as Google, Instagram, Spotify, Netflix, Uber, Dropbox, Pinterest and Reddit.

Wrapping code

It is not always possible or useful to rewrite the entire code of existing FORTRAN or MATLAB models in Python for obvious reasons, only to make it more accessible for potential others. This process of line-to-line translation might take months and could only be taken up by a researcher who is familiar with the underlying logic of the model in question and in both languages, FORTRAN and Python. Nevertheless, it may often be desirable or even necessary to preserve existent models programmed in (what we may call) esoteric and legacy computer languages[59] and make them compatible with contemporary software

59 It is important to make the difference here between legacy and esoteric programming languages. By 'legacy languages,' I understand programming languages that are no longer in use today, partly because of their incompatibility with current technical systems. By 'esoteric languages,' in contrast, I mean that only a few people have an expertise in programming them. Not all systems can be programmed in Python or other flexible languages. The

technologies and infrastructures. Developers can develop *wrappers,* which translate code from one language into another, to do so. An example of this tactic is *Pymagicc,* a Python interface for the FORTRAN-based climate model MAGICC (Model for the Assessment of Greenhouse-Gas Induced Climate Change; Meinshausen et al. 2011). The original MAGICC model is used by several modeling groups to assess the pathways of future emissions in climate policy analyses. The promise of Pymagicc is that it mobilizes MAGICC for actors that are not familiar with FORTRAN and the model-specific software environment. By contrast, Pymagicc runs on Windows, macOS and Linux and uses a standard tabular data structure (DataFrames from the Pandas library) for emissions scenarios. As a result, the MAGICC model parameters and emissions scenarios can be modified using Python, without having to touch the original FORTRAN model. Considering that such Python wrappers have recently been developed for a number of climate models, this practice also provides new opportunities for model comparison. The Pymagicc source code, documentation, an issue tracker and Jupyter Notebook are made available via a Github repository (see further below). The model can even be explored interactively in a web browser via the Binder project[60] (Gieseke et al. 2018: 1f).

Tapping Crowdknowledge

The use of Python by industrial actors also points to another ground for Python's increasing dominance: By choosing a particular entanglement

contemporary Machine Learning algorithms are all programmed using precise languages, such as C, rather than Python. But they typically come with a Python wrapper, which allows them to be accessible to a wider user community. An example is Googles' machine learning library Tensorflow, whose core runs on highly optimized C++, while providing direct manipulation via Python.

60 https://mybinder.org/v2/gh/openclimatedata/pymagicc/master?filepath=notebooks/Example.ipynb.

of technologies (e.g. Python, common packages) and way of doing things (open sourcing code), scientists become part of an ever-exploding community of practice (Lave 1991) gathering around terms such as 'data science,' 'data analytics' and 'machine learning.' As VanderPlas highlights, "[...] if there is a scientific or data analysis task you want to perform, chances are someone has written a package that will do it for you" (2016: 2). On the one hand, this means that developers will easily find ready-to-use code snippets solving all sorts of problems encountered by others which can be integrated into ones' own programming code. On the other hand, the web and its dedicated platforms provide millions of troubleshooting tips for coding issues. While there are multiple platforms on the web providing such services, the most prominent ones are github.com, stackoverflow.com and medium.com.

Github is an American company and community platform providing a variety of services for software development, including hosting, distributed version control (Git[61]) and source code management. It also provides access control and several collaboration features, such as bug tracking, feature requests, task management and wikis for every project. Github is a subsidiary of Microsoft, which acquired the company in 2018. GitHub offers plans for free, and professional and enterprise accounts, and its free accounts are commonly used to host open-source projects. We can take the example of the CLIMADA profile (see Fig. 35) to describe the structure of content on Github. The main view of a repository[62] on Github shows the project title, buttons for community functions (watch/star/fork), a number of content structuring menus (code, issues, pull requests, projects, wiki, security and insights), statistics (1249 commits, 2 branches, 14 releases, 10 contributors, license info), the file and folder structure of the project, and the text of the

61 More info on https://git-scm.com/, retrieved on June 2, 2019.

62 https://github.com/CLIMADA-project/climada_python, retrieved on June 2, 2019.

readme file (including installation instructions and project documentation. One can easily download the packaged code ('clone or download') or contribute to the project (e.g. 'create new file,' 'upload files').

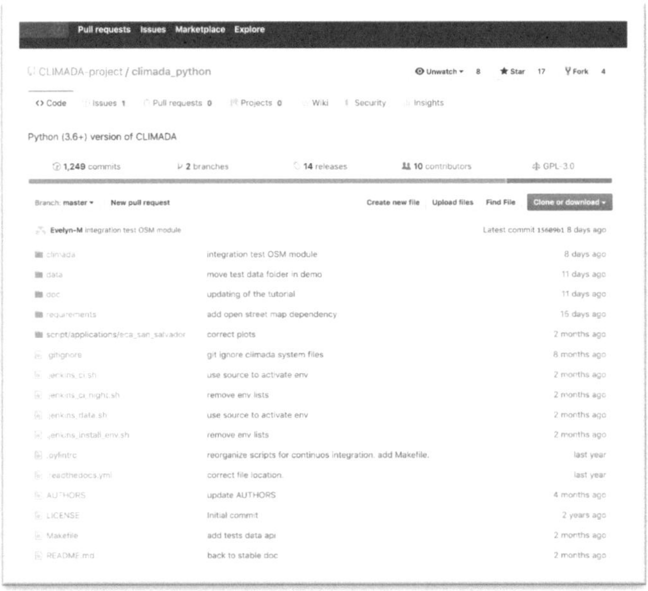

Figure 35: View of the CLIMADA repository[63] on Github.
Source: My own screenshot

On a technical level, the key feature of Github is the underlying technology *Git*, a free and open-source distributed control system version, which is commonly used nowadays in collaborative software development. With Git, every change to a software code becomes perfectly accountable and traceable for others. 'Others' here can mean the members of a well-defined and closed community[64] (e.g. a company, network or

63 https://github.com/CLIMADA-project/climada_python, retrieved on June 5, 2019.

64 E.g. through deployments of Git, such as gitlab. See http://about.gitlab.com, retrieved on June 5, 2019.

organization) or – as in the case of Github – the World Wide Web. Provided the legal preconditions (an open software license) have been met, developers can also create ('fork') and develop their own version of an existing software project on Github, making this provenance and genealogy perfectly visible and traceable. On the other hand, Github is essentially a social network connecting developers and users of software projects. Similar to Stack Overflow, Github enables software developers to publicly ask questions related to programming code and infrastructure. In contrast to Stack Overflow (see below), the discussions are happening around a particular project, its developers and community.

Stack Overflow is a privately-held community platform for programmers and businesses, featuring questions and answers on a wide range of topics in computer programming. Users find these answers either by entering questions into the search console of the platform or by Googling it and being forwarded to the site. The answers are rated by the community for their usefulness and ranked accordingly by the platform. In the rare cases where an answer to a question is not available in the archive, one can open a new thread and ask the community for help. As a result, stack Overflow facilitates a nearly perfect information clearance between questions and answers to programming issues. The more common a problem, programming language and technological stack (e.g. data issues in Python), the greater the chance that an answer can be found on Stack Overflow.

While Github gathers actors around a specific code and Stack Overflow provides solutions to concrete coding problems, *medium.com* is the platform to negotiate more conceptual issues linked to software development. Medium is an online publishing venture launched in August 2012. The Wikipedia article of the platform describes it as "an example of social journalism, having a hybrid collection of amateur and professional people and publications, or exclusive blogs or publishers on

Medium, and is regularly regarded as a blog host."[65] Initiated by a co-founder of Twitter (Evan Williams), the original idea for Medium was to provide a way to publish writing and documents longer than Twitter's 140-character (now 280-character) maximum. Its self-image is organized around promises of innovation, fresh ideas and creativeness:

> Ideas and perspectives you won't find anywhere else.
> Medium taps into the brains of the world's most insightful writers, thinkers, and storytellers to bring you the smartest takes on topics that matter. So whatever your interest, you can always find fresh thinking and unique perspectives.[66]

Issues discussed on Medium are not limited to technology but the software development and startup communities are among the most active ones on the platform. Entering 'why learn python' in Medium's search console returns a multitude of articles, one published within the influential social blog *Hackernoon* titled *10 Reasons to Learn Python in 2019*. The items on the list confirm many of the aspects discussed within this chapter and include "Data science, Machine Learning, Web development, Simplicity, Huge community, Libraries and frameworks, Automation, Multipurpose, Jobs and Growth, and Salary."[67]

With the rise of community platforms, such as Github, Stack Overflow and Medium, programming code and programming knowledge has become increasingly distributed between different spaces, actors and artifacts. As Adrian Mackenzie has highlighted earlier, "[...] software has hybridized itself wildly with other media and practices and is likely

65 See Wikipedia's entry for medium.com at https://en.wikipedia.org/wiki/Medium_(website), retrieved on April 2, 2019.

66 See https://medium.com/about, retrieved on April 2, 2019.

67 See https://hackernoon.com/10-reasons-to-learn-python-in-2018-f473dc35e2ee, retrieved on April 2, 2019.

to continue doing so [...]" (2006: 9). This hybridization has only truly kicked in within science very recently.

Mobile calculation and accountability

A very specific technology about to change contemporary scientific programming is the *Jupyter notebook,* a hybrid device between interactive computational environment and scientific documentation. Influenced by existing projects, such as Mathematica's notebook and IPython (Interactive Python), the notebooks were designed to "support the workflow of scientific computing, from interactive exploration to publishing a detailed record of computation" (Kluyver et al. 2016: 88). Considering that CLIMADA is delivered as Jupyter Notebook, we can use it to describe the functionalities of this technological device. The appearance of the notebooks is inconspicuous and discreet. They basically present themselves as simple HTML websites to be displayed in any web browser. This visual appearance is a stark understatement, as the whole fairly complex CLIMADA model is visible, operational and modifiable from within the notebook. A browser window is opened showing the file structure of the model when Jupyter and CLIMADA are initiated on a laptop.[68] One can browse within the folder and file structure, showing all elements of the CLIMADA programming code.

In our case, the data of these files are stored locally on a Macbook Pro laptop. However, the same setting would be equally deployable on a distributed cloud-computing environment,[69] thereby enabling more intensive operations computationally, such as machine learning. Operations within CLIMADA can be undertaken from within the Notebook

68 Via the Integrated Development Environment (IDE) Anaconda. https://www.anaconda.com/, retrieved on April 3, 2019.

69 E.g. an Amazon Elastic Compute Cloud (EC2). See https://docs.aws.amazon.com/dlami/latest/devguide/setup-jupyter.html, retrieved on July 5, 2019.

files within the folder structure, in our case '1_main_climada.ipynb,' shown in Figure 36. The element in focus is the calculation of 'exposures' within CLIMADA.

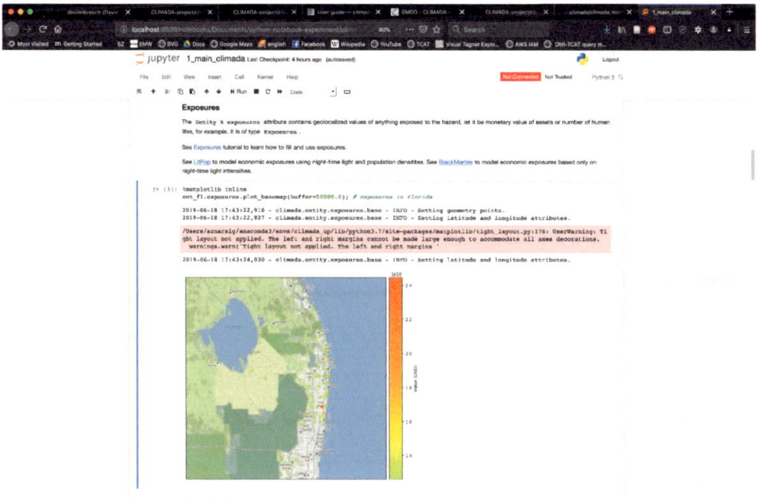

Figure 36: One of CLIMADAs Jupyter notebooks, the file '1_main_climada.ipynb.' Source: Own screenshot

The explanation of exposures says:

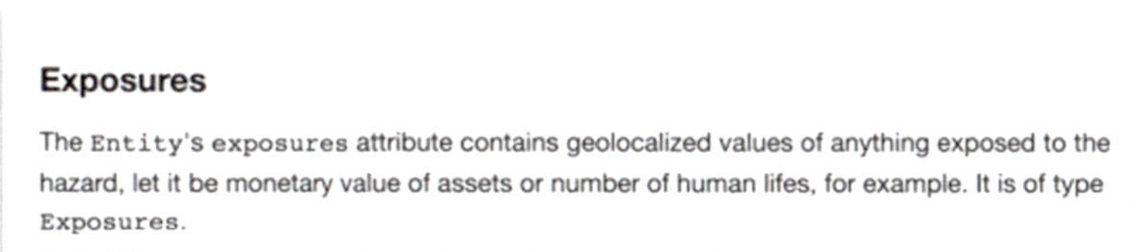

Exposures

The `Entity`'s `exposures` attribute contains geolocalized values of anything exposed to the hazard, let it be monetary value of assets or number of human lifes, for example. It is of type `Exposures`.

This general explanation is followed by a link to a tutorial (another notebook) that helps one to "learn how to fill and use exposures." The notebook then links to two specific submodels (or classes), 'LitPop' and 'BlackMarble,' which are alternative methods to model exposure to natural hazards.

> See Exposures tutorial to learn how to fill and use exposures.
>
> See LitPop to model economic exposures using night-time light and population densitites. See BlackMarble to model economic exposures based only on night-time light intensities.

The notebooks of LitPop and BlackMarble are both connected to the main CLIMADA notebook. Therefore, one can change elements within BlackMarble, which will be taken into account within any further CLIMADA calculation. It then follows a grayed-out text box, which highlights the calculative element of this notebook block:

```
%matplotlib inline
ent_fl.exposures.plot_basemap(buffer=50000.0); # exposures in Florida
```

This is a command line written in Python. It tells CLIMADA to plot exposures according to functions defined earlier in the text of the notebook. If one presses 'ALT + ENTER' on the keyboard, the notebook will calculate the exposures to a natural hazard in the territory of Florida and plot the results as dots in false colors on a map of the US state. The visual representation is enabled by calling up 'matplotlib,' a standard Python package for visualization purposes. In this particular case, the calculation takes less than one second.

The notebook then acts as a log file documenting the calculations performed by the machine (in this case, the CPU of my laptop):

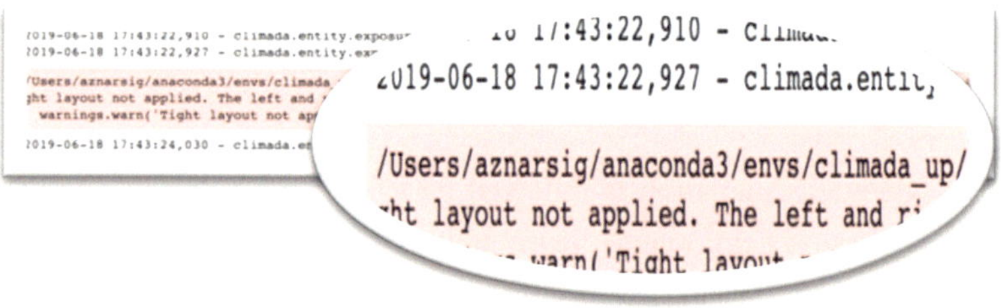

The notebook does not only document calculations performed successfully but can return concrete error messages. In the present case, it informs us about a slight representational issue on the map: "Tight layout not applied. The left and right margins cannot be made large enough to accommodate all axes decorations." In so doing, Jupyter also helps with troubleshooting in the interactive coding process.

The Jupyter notebook draws together many functionalities that are traditionally distributed among various technologies and artefacts, including:

- running code, such as a shell and command-line interface;
- organizing file structures and computational environments, such as an operating system;
- providing access to the entire programming code, as in a code repository;
- writing programming code, as in a code editor;
- producing and displaying laid out diagrams and formatted text, as in word processors and design software;
- documenting the scientific process, results and methodology, as in software user manuals and methods chapters of a publication;
- enabling data exploration and analysis, similar to proprietary software for visual analytics;
- providing a procedural tool for scientists to structure their work, similar to notes, post-its and various organization software;
- facilitating the collaboration in spatially distributed teams using cloud computing environments;
- facilitating replicability and reuse; and
- enabling various forms of monitoring and evaluation.

The Jupyter Notebook draws these functionalities together in one place, which can be easily accessed via a web browser. It provides an overview and possibilities for the manipulation and control of other distributed and fluid elements, such as data, programming code, computation, visualization and documentation. Within the daily programming practice, it is irrelevant whether these elements are stored on and retrieved from the physical computer located in front of the researcher or based on a distributed cloud computing infrastructure, such as the PIK's supercomputer, or Amazon's AWS.[70] As a matter of fact, commercial cloud computing services, such as AWS and Google Cloud, provide detailed instructions on how to set up Jupyter Notebooks within their infrastructures.[71]

Mobile modeling

I propose to subsume the discussed entanglement of practices and infrastructures under the term *mobile modeling*. Mobile modeling in science aims at profiting from the power of today's distributed computing technologies and infrastructures. Mobile modelers use the same programming languages and rely on similar software packages as programmers within the global knowledge economy. As a result, they are able to benefit from a powerful community of practice around data analysis ('data-science') from its continuously optimized software stacks and its powerful cloud computing infrastructures. The strategies of mobilization discussed above (pythonizing, packaging, wrapping code, tapping distributed communities) amplify the possibilities of scientific modeling endeavors. However, they also create considerable challenges for

70 Amazon Web Services.

71 For a detailed instruction in AWS, see https://docs.aws.amazon.com/dlami/latest/devguide/setup-jupyter.html, retrieved on April 3, 2019.

trust, control, performance and scientific soundness. Packages such as MatPlotLib are frequently updated to ensure the performance within a fluid environment of gradually evolving software stacks and infrastructural entanglements. Regarding scientific programming, this means that experiments have often been carried out with versions of software packages that might no longer be operational or available. Mobile modelers are trying to stabilize the increasingly distributed and fluid elements of their modeling environment with devices such as the Jupyter Notebook in order to keep some control over the scientific process. It is not surprising that the Jupyter notebook is presented by its developers as the Swiss army knife for future computational and data science. It appears like a technological fix to the perceived reproducibility crisis in science (Baker 2016; Kitchin 2014a; Marwick 2015), which is vividly debated in fields such as climate impact research.

Mobile modeling is essentially a forward-looking practice concerned about the expected mobility of modeling technology in the future, paralleled with strategies for prospective stabilization (also see the discussion in the next chapter). It always gives preferential treatment to technologies and infrastructures with capabilities of dealing with the distributiveness and fluidity of components.

V. Future data

How does one describe data and the work they do? How does one get access to data and strategies for their characterization? In an interview with simulation modeler Jana Solberg, she tells me about her work as a member of the modeling team generating the SSPs. The purpose of the latter is to serve as a standardized set of socioeconomic storylines that can 'drive' all global climate-modeling endeavors in the world. When Jana was talking about the SSPs, she pointed at elements displayed on the website, data repository and viewer *SSP database* – file and folder structures, diagrams and textual descriptions. After the interview, I come back to the database to learn more about the SSPs and their representation as digital datasets.

Access log

Based on the information obtained during my interview with Jana, I type 'ssp database' into Google's query tool[72] and click on the first of the search results, forwarding me to a bulky URL.[73] The main dashboard is hidden by a pop-up window displaying an agreement to Terms of Use. It includes terms on copyrights, citations required and liabilities. The agreement aims at protecting the provider of the online repository (here IIASA, the International Institute for

72 Or rather www.qwant.com or www.startpage.com, retrieved on April 3, 2019.

73 https://secure.iiasa.ac.at/web-apps/ene/SspDb/dsd?Action=htmlpage&page=about, retrieved on April 3, 2019.

Applied Systems Analysis) and the developers of the datasets (e.g. the PIK). Clicking 'I agree to the Terms of Use,' a next checkpoint is waiting for me: "Please use this button to log in as a guest user (restricted preview) or use the form below (and your individual email and password) to log in with your email." Doing the latter, I finally receive access to the website. The design of the website is very simple, with links to subpages such as 'welcome,' 'basic elements,' 'IAM scenarios,' 'CMIP6 Emissions', 'download' and 'citation.' I browse through the content and get caught by 'IAM scenarios,' identifying it as the cornerstone of the website. The subpage is structured as a so-called data viewer, representing the datasets of the SSPs in various forms (see Figure 37).

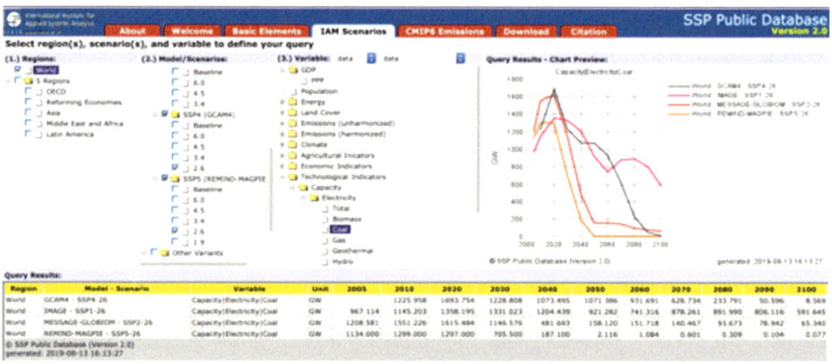

Figure 37: Data viewer of the SSP database. Source: IIASA Website

I try to figure out the structure and functionality of the data viewer. It becomes clear to me while clicking through the elements that the content is structured from the top left to the bottom right, similarly to a book's page. Based on prior knowledge gathered through the field research, I click through the structure choosing the region 'world,' four different SSPs, a low scenario for future GHG emissions (RCP 2.6) and one particular simulated element (coal electricity). The diagram on the right now displays the data outputs of the simulations according to the choices made.

```
The x-axis represents time (years 2000 to 2100) and the y-
axis the coal electricity capacity (gigawatt). I look at the
diagram and try to interpret the form of the lines: If the
world wants to limit global warming to 2 degrees (RCP 2.6),
coal electricity should decrease as early as in the year
2020, independently of the prospect of the economy and other
factors within the next years. I feel happy. After four years
of my PhD research, I finally managed to grasp the basics
of the SSPs to infer one major insight of the modeling work
and to articulate it within this text. A little proud of
this successful simulation of my informants' practices, I
make a screenshot of the website and integrate it as an image
above.
```

In a similar way as the models and code in chapter IV, we will now investigate how scientists make data and their information travel across contexts. On the one hand, this may help to obtain a better pragmatic understanding of 'research data.' On the other hand, it aims at characterizing the current transformation towards open data, infrastructures and services.

About data

There are not many terms in the contemporary world as omnipresent and problematic as 'data.' As Lisa Gitelman and Virginia Jackson have put it in the introduction of the anthology *'Raw Data' Is an Oxymoron*:

> Data are everywhere and piling up in dizzying amounts. Not too long ago storage and transmission media helped people grapple with kilobytes and megabytes, but today's databases and data backbones daily handle not just terabytes but petabytes of information, where *peta-* is a prefix which denotes the unfathomable quantity of a quadrillion, or a thousand trillion. (2013: 1)

However, data are not just *there,* they also *do* things with us and our world.

> Data are units or morsels of information that in aggregate form the bedrock of modern policy decisions by government and nongovernmental authorities. Data underlie the protocols of public health and medical practice, and data undergird the investment strategies and derivative instruments of finance capital. Data inform what we know about the universe, and they help indicate what is happening to the earth's climate. (ibid.)

This last reference to the Earth's climate points us to another crucial aspect of data. It is not only understood to give us information about the past and the present but also about our future. As Gitelman and Jackson quote from a famous IBM advertisement: "Our data isn't just telling us what's going on in the world, it's actually telling us where the world is going" (IBM cited in ibid.).

Ignoring the particular production context of this slogan (advertisement), this statement may actually be true and meaningful. In the particular case of climate research, the translation of GHGs and temperatures into handy datasets has not only enabled science to identify and prove a long-term trend toward global warming, it also provides a representational method of making statements about probable continuations of this trend in the future. Investigating 'data' in the context of climate impact research is interesting for a number of reasons. As a matter of fact, these researchers are engaging many things that have been problematized in other contexts: They have dealt almost exclusively with massive amounts of data ('Big Data'?) for a long time, engaged in complex analytic activities based on this data ('data science'?) and made predictions that have an impact on the lives of others (predictive analytics?). Without anticipating the following analysis in this chapter, we may argue that climate impact modelers do many of the things that are currently hyped and problematized elsewhere, but they do it a bit differently. Making use of a figure of speech introduced by Geoffrey Bowker, they seem to cook data with a lot of care (2005: 194). How may we characterize data? Rob Kitchin's seminal book *The Data Revolution* starts with the following preliminary description:

> Data are commonly understood to be the raw material produced by abstracting the world into categories, measures and other representational forms – numbers, characters, symbols, images, sounds, electromagnetic waves, bits – that constitute the building blocks from which information and knowledge are created. (2014b: 1)

The representative aspect of data for Kitchen is not always be explicit but may also be implied or derived:

> Data are usually representative in nature (e.g. measurements of a phenomena, such as a person's age, height, colour, blood pressure, opinion, habits, location, etc.), but can also be implied (e.g., through an absence rather than presence) or derived (e.g., data that are produced from other data, such as percentage change over time calculated by comparing data from two time periods) [...]. (ibid.)

Data can either be recorded and stored in analog form or encoded as binary digits (bits). It may be categorized by form (qualitative or quantitative), structure (structured, semi-structured or unstructured), source (captured, derived, exhausted, transient), producer (primary, secondary, tertiary) and/or type (indexical, attribute, metadata) (ibid.: 4ff).

About research data

How can we get a grasp of scientific data and of the work it does? If we understand knowledge as situatively produced entities (Haraway 1988), it appears meaningless to characterize 'data' here in the abstract.

Figure 38: Soil stored in the pedocomparator transformed into an inscription. Source: Latour (1999a: 55)

Bruno Latour has addressed the construction and mobilization of scientific data in his article on *Circulating Reference* (1999a), tracing the transformation of the Amazon rainforest into botanic specimens, soil samples, tables, maps and, finally, into an academic publication. Building on this detailed description of transformative processes in science, we may ask: When does the forest cease to be forest and become 'data'? We should be careful of making this a categorical shift in science, but we can identify one instant that appears crucial in this becoming of data, namely, when soil samples stored and arranged in the instrument of the 'pedocomparator' are transformed into diagrammatic form on a piece of paper (see Figure 39): "We move now from the instrument to the diagram, from the hybrid earth/sign/drawer to paper" (ibid.: 54). This transformation is what Latour refers to as *inscription:*

> A general term that refers to all the types of transformations through which an entity becomes materialized into a sign, an archive, a document, a piece of paper, a trace. (Latour 1999b: 306)

In fact, 'inscription' for Latour is not only the process but also the resulting artifact:

> Usually, but not always inscriptions are two dimensional, superimposable, and combinable. They are always mobile that is, they allow new translations and articulations while keeping some types of relations intact. (ibid.: 306f)

This is where Latour equates 'inscription' with another of his concepts, the *immutable mobile*, "[...] a term that focuses on the movement of displacement and the contradictory requirements of the task" (ibid.:307). While we might contend that the 'inscription' and 'immutable mobile' oscillate well with concepts of 'data' (see Rheinberger's interpretation below), Latour's text is actually rather unspecific in his use of the term:

> [...] [A]n enormous pile of newspaper stuffed with plants brought back from the site and awaiting classification. The botanist has fallen behind. It is the same story in every laboratory. As soon as we go into the field or turn on an instrument, we find ourselves drowning in a sea of data. (I too have that problem, being incapable of saying all that can be said about a field trip that took only fifteen days.) Darwin moved out of his house soon after his voyage, pursued by treasure chests of data that ceaselessly arrived from the *Beagle*. Within the botanist's collection, the forest, reduced to its simplest expression, can quickly become as thick as the tangle of branches from which we started. (Latour 1999a: 39)

Therefore, 'data' is equally associated with specimens stuffed in newspapers, nonspecified material on Darwin's ship and, more broadly, with what we might call 'overwhelming impressions from the ethnographic field.' Latour shows a specific interest in the term 'data' only in one instance, highlighting its problematic etymology:

> In order for the botanical and pedological data to be superposed on the same diagram later, these two bodies of reference must be compatible. One should

> never speak of "data" – what is given – but rather of *sublata,* that is, of 'achievements.' (ibid.: 42)

This critique oscillates with arguments made by many others (Bowker 2005; Gitelman 2013; Kitchin 2014b; Leonelli 2015), which will be discussed more in detail further below. While Latour's article has not engaged in a fundamental characterization of data, it certainly influenced conceptualizations that followed. Hans-Jörg Rheinberger draws on Latour's arguments in his article *Infra-Experimentality*, translating the example of soil samples to genetics and the practice of genome sequencing:

> To stay with our molecular example, a next step consists in transforming the sequence gel into a chain of symbols standing for the four nucleic acid bases. With this visual display total abstraction is made not only from the particle from which the nucleic acid was extracted, but also from the test tube reaction in which it was differentially synthesized, and moreover from the gel and its material qualities in which the fragments were separated. (Rheinberger 211: 343)

In a similar way as Latour transforms soil samples in the pedocomparator to a map on paper, the sequence gel (trace) is transformed into a chain of symbols (data) containing the *information* for the expression of a protein. This is, according to Rheinberger, where "traces" become "data":

> The most important thing perhaps in such transitions is: the result of the experiment is brought into a form in which it can be *stored*, and consequently, *retrieved* as well. Much speaks for the assumption that the ability to be stored, that is, to be made *durable*, is the most important prerequisite for transforming *traces* into *data*. (ibid., emphasis in the original)

For Rheinberger, this is the shift in which immutable mobiles are born:

> Traces are not, but data are of the form of Latourian "immutable mobiles". Their relative immutability is a prerequisite for their mobility, their retrievability, their options for becoming re-enacted, and all the rest we associate with data and not with – usually precarious, bound-to-disappear – traces. (ibid., 344, emphasis in the original)

In this reading, data are synonymous with inscriptions and immutable mobiles. Data emerge in the moment when all material traces are exchanged against pure symbolic inscription. Both the articles by Rheinberger and Latour evoke the question to what extent the proposed characteristics of data, inscriptions and immutable mobiles are universal or specific to the sphere of research and the natural sciences in particular. This is especially the case for Latour, who has discussed the work of immutable mobiles more independently from science in his article on *Visualization and Cognition: Drawing Things Together* (Latour 1988). The text describes immutable mobiles in the form of cartographic inscriptions, which transformed the knowledge of power relations in colonial settings considerably. While maps may be produced by means of scientific instruments, the setting examined is clearly not one of science. Latour's work on immutable mobiles is generally so productive not only because it can stand for scientific knowledge production but also for knowledge production in general. The immutable mobile concept is itself an immutable mobile, traveling through the worlds of philosophy, the history of science and technology, and the sociology of knowledge. This consciously constructed vagueness of scale exists equally for the concept of Latour's *circulating reference:* "When immutable mobiles are cleverly aligned they produce the circulating reference" (Latour 1999b: 307) Latour is rather unspecific regarding in what reference frame this circulating reference is operating – as a philosophical category for 'sense-making in the world' or rather as a description of 'sense-making in science.'[74]

Data as relational property

Science studies scholar Sabina Leonelli builds on Rheinberger's and Latour's arguments but also criticizes their universal claims regarding data (inscriptions or immutable mobiles). The question for Leonelli what data is cannot be answered only by assessing its material qualities (mobility, stability across contexts) and degree of manipulation (inscription into symbolic form). Rather, she understands data as a purely relational property that can only be identified with reference to concrete research situations and the decisions and perceptions involved:

> A better option is give up altogether on a definition of data based on the degree to which they are manipulated, and focus instead on the relation between researchers' perceptions of what counts as data and the stages and contexts of investigation in which such perceptions emerge. (Leonelli 2015: 5)

In my opinion, this understanding of data has some strong argumentative points. Compared to Rheinberger's characterization, it diminishes the subliminal bias towards data produced within the natural sciences. While Rheinberger focuses on highly structured, numerical datasets, Leonelli's perspective may be better suited to incorporate the being of unstructured, heterogeneous data, for example, from ethnographic research or variations of web- and data-science: "Data can therefore

74 Bruno Latour only makes explicitly clear that the description discards the perspective of sociology: "Of course had I not artificially severed the philosophy from the sociology, I would have to account for this division of labor between French and Brazilians, mestizos and Indians, and I would have to explain the male and female distributions of roles" (Latour 1999a: 44).

include experimental results as well as field observations, samples of organic materials, results of simulations and mathematical modeling, even specimens." (ibid.: 6)

Compared to Bruno Latour's immutable mobiles, Leonelli gives more weight to the prospective and perceptional aspects of data. From this perspective, the material form of artifacts is not the only constitutive feature of data. Equally, data must have been collected, stored and disseminated with the expectation of being used as evidence for knowledge claims. This does not necessarily mean that scientists know *how* the data might be used in the future (ibid.). This is an important point from an epistemological perspective and a critique of common understandings of data as "numbers, characters or images that designate an attribute of a phenomenon" (Royal Society cited in ibid.: 7). For her, this fundamental link between data and phenomena is not given. On the one hand, "researchers often produce data without knowing exactly which phenomenon they may document" (ibid.: 6). Researchers may produce data because they have access to particular instruments and they hope that it might later be helpful to identify new, unknown phenomena. On the other hand, the same data may act as evidence for a variety of phenomena, depending on the situational context (ibid.). However, similar to Latour and Rheinberger, Leonelli acknowledges the aspect of the *portability* of data as a precondition for its use as evidence:

> No intellectual achievement, no matter how revolutionary and well-justified, can be sanctioned as a contribution to scientific knowledge unless the individual concerned can express her ideas in a way that is intelligible to a community of peers, and can produce evidence that can be exhibited to others as corroborating her claims. [...] If data are not portable, it is not possible to pass them around a group of individuals who can review their significance and bear witness to their scientific value. (ibid.: 7)

The characterization of data as 'portable objects' draws from common terminology in computer science, where software portability is understood as “a property of a program that can run on more than one kind of computer” (Downey 2012: 7). Equally, high-level languages (see Chapter IV) are portable, “meaning that they can run on different kinds of computers with few or no modifications. Low-level programs can run on only one kind of computer and have to be rewritten to run on another” (ibid.: 1). The term ‘data portability’ has only recently gained momentum in the context of the fight for digital rights. The recently established European General Data Protection Regulation, for example, includes the “right to data portability” (Article 20):

> The data subject shall have the right to receive the personal data concerning him or her, which he or she has provided to a controller, in a structured, commonly used and machine-readable format and have the right to transmit those data to another controller without hindrance from the controller to which the personal data have been provided [...].[75]

Data portability is closely related to the discourse of open data, which is introduced as follows.

Open data

In the past, access to valuable data and information has traditionally been restricted in some ways, for example, through financial, legal, organizational, technical and/or cognitive barriers. Accessing an academic publication, for example, might require the payment of a fee (financial) to a publisher in order to obtain the right (legal) to download an article. Equally, access to the publication might be restricted because it is only stored at one geographical location, in a state archive and in paper form (organizational). Vice versa, an exclusive availability online and in digital form creates new technical and cognitive barriers, as the discussion about the *digital divide* is showing (Norris 2001).

75 https://gdpr-info.eu/art-20-gdpr/, retrieved on April 16, 2019.

Against this backdrop, the open data movement seeks to change this situation radically, making data potentially available to anyone.

As Rob Kitchin highlights, the movement is built on three principles: Openness, participation and collaboration. "Its aim is to democratize the ability to produce information and knowledge, rather than confining the power of data to its producers and those in a position to pay for access" (Kitchin 2014b: 48). On the one hand, attention has been focused on opening up public data emanating from state authorities and from research institutes, given that these have been funded by the public purse for the public's benefit. On the other hand, open data is also increasingly being pushed by the private industry, with anticipations of an innovative push through such practices. The open data community is interwoven with other movements fighting for the right to information, open knowledge, open-source software and open science. Within the last century, 'open data' has become increasingly popularized and mainstreamed through media campaigns (e.g. The Guardian's *Free Our Data*[76]), the call and endorsement of open data policies by inter- and supranational organizations (e.g. Organization for Economic Cooperation and Development[77] and the European Union[78]), national governments (e.g. Germany[79]) and municipal authorities (e.g. Berlin[80]) (ibid.).

Open Research Data

The idea of open public data had to be translated to the particularities of research practice to be meaningful for the case of science. This translation has recently given way to the *FAIR data principles*, a number of

76 http://www.freeourdata.org.uk/, retrieved on April 5, 2019.

77 https://data.oecd.org/, retrieved on April 5, 2019.

78 https://data.europa.eu/euodp/en/data/, retrieved on April 5, 2019.

79 https://www.govdata.de/, retrieved on April 5, 2019.

80 https://daten.berlin.de/, retrieved on April 5, 2019.

guidelines supported and endorsed by various organizations in science and beyond:

Box 2 | The FAIR Guiding Principles

To be Findable:
F1. (meta)data are assigned a globally unique and persistent identifier
F2. data are described with rich metadata (defined by R1 below)
F3. metadata clearly and explicitly include the identifier of the data it describes
F4. (meta)data are registered or indexed in a searchable resource

To be Accessible:
A1. (meta)data are retrievable by their identifier using a standardized communications protocol
A1.1 the protocol is open, free, and universally implementable
A1.2 the protocol allows for an authentication and authorization procedure, where necessary
A2. metadata are accessible, even when the data are no longer available

To be Interoperable:
I1. (meta)data use a formal, accessible, shared, and broadly applicable language for knowledge representation.
I2. (meta)data use vocabularies that follow FAIR principles
I3. (meta)data include qualified references to other (meta)data

To be Reusable:
R1. meta(data) are richly described with a plurality of accurate and relevant attributes
R1.1. (meta)data are released with a clear and accessible data usage license
R1.2. (meta)data are associated with detailed provenance
R1.3. (meta)data meet domain-relevant community standards

> There is an urgent need to improve the infrastructure supporting the reuse of scholarly data. A diverse set of stakeholders – representing academia, industry, funding agencies, and scholarly publishers – have come together to design and jointly endorse a concise and measurable set of principles that we refer to as the FAIR Data Principles. (Wilkinson et al. 2016: 1)

Table 2: The FAIR Guiding Principles.
Source: Wilkinson et al. (2016: 4)

'FAIR' denotates four different qualities of data: 'Findable,' 'Accessible,' 'Inter-Operable' and 'Re-usable' (see Table 2). Rather than discussing these principles in detail, we will now consider concrete open data practices in the field of climate impact research. As we will see, these practices oscillate with the aspects of knowledge mobilization discussed by Latour, Rheinberger and Leonelli.

Open data in action

We have already come across Tobias Geigers' work on extreme weather events and the simulation of their economic damage. Figure 39 shows a snapshot of an animation that Geiger uses to describe and explain his model and simulation, in this case, within my interview with him at the Potsdam Institute. The animation shows a dynamic simulation of damage from hurricanes within the territory of Bangladesh. It essentially works and aesthetically looks like a missile trajectory simulation.

Figure 39: Animation of damage from hurricanes within the territory of Bangladesh. Source: https://vimeo.com/user49173690 by David Bresch, filmed during interview with Tobias Geiger

In the following, we will take a closer look at one particular dataset produced by Geiger and other scientists: Spatially-explicit Gross Cell Product (GCP) time series: past observations (1850–2000) harmonized with future projections according to the Shared Socioeconomic Pathways (2010–2100) (Geiger et al. 2017). The data consists of values for GDP in a temporal series spatial grid and temporal time series. The dataset has been used as input data for a number of scientific projects, such as the simulation of past and future damages of hurricanes modeled in/with CLIMADA (see Chapter IV), which is maintained by researchers at the Swiss ETHZ. It also builds on other work carried out at the Potsdam Institute, namely the SSPs as scenarios for future socio-economic development. The dataset is stored on a server maintained by the shared library services of the research institutes on Telegrafenberg. It can be accessed via a library information sheet, similar to traditional academic publications. The description on the information sheet reads as follows:

> We here provide spatially-explicit economic time series for Gross Cell Product (GCP) with global coverage in 10-year increments between 1850 and 2100 with a spatial resolution of 5 arcmin. GCP is based on a statistical downscaling procedure that among other predictors uses national Gross Domestic Product (GDP) time series and gridded population estimates as input. Historical estimates until 2000 are harmonized with future socioeconomic projections from the Shared Socioeconomic Pathways (SSPs) according to SSP2 from 2010 onwards.
>
> We further provide a mapping file with identical spatial resolution to associate GCP values with specific countries. Based on this mapping we provide nationally aggregated GDP estimates between 1850–2100 in a separate csv-file.
>
> Additionally, we provide a mapping file with identical spatial resolution providing national assets-GDP ratios, that can be used to transform GCP to asset values based on 2016 estimates from Credit Suisse's Global Wealth Databook 2016.[81]

The terms of use of the dataset are specified as CC BY 4.0, a creative commons license allowing for sharing (copy and redistribute the material in any medium or format) and adapting (remix, transform and build upon the material) for any purpose, even commercial. However, the

81 http://dataservices.gfz-potsdam.de/pik/showshort.php?id=escidoc:2740907, retrieved on April 5, 2019.

prospective user must give appropriate credit, provide a link to the license and indicate if changes were made. Equally, he/she is not allowed to apply legal terms or technological measures that legally restrict others from doing anything the license permits (Creative Commons 2019). In legal terms, this is the constitutional element to make this *open* data. The set includes four files (here with the description from the library page) in a zip-folder:

- GCP_PPP-2005_1850-2100.nc: GCP in 10-year increments between 1850 and 2100 with a resolution of 5 arcmin.
- National_GDP_PPP-2005_1850-2100.csv: nationally-aggregated GDP estimates (as used for GCP downscaling) in 10-year increments between 1850 and 2100.
- ISO-country-map.nc: Map for grid cell to ISO 3166 country code mapping with a resolution of 5 arcmin.
- GDP2Asset_converter_5arcmin.nc: Map for grid cell GDP to Asset mapping with a resolution of 5 arcmin based on 2016 estimates from Credit Suisse's Global Wealth Databook 2016.[82]

The four files are what my interview partners at PIK often referred to as 'raw data.' STS scholars, in contrast, have problematized this term on various occasions. The most prominent example is Geoffrey Bowker, who declared: "Raw data is both an oxymoron and a bad idea; to the contrary, data should be cooked with care" (2005: 184) This idiom later inspired the anthology *Raw Data is an Oxymoron* edited by Lisa Gitelman (2013). The contributors in the anthology open various perspectives of the investigation of data. All articles are driven by the belief that we should refrain from considering 'data' in its etymological sense

82 http://dataservices.gfz-potsdam.de/pik/showshort.php?id=escidoc:2740907, retrieved on April 5, 2019.

as given. Instead, we should carefully assess how data are 'cooked' within different social settings, circumstances and technological entanglements. In other words, data have history and this history matters. Against this background, a number of authors have introduced alternative terms to 'data,' considering their active generation. Exemplary are Bruno Latour's "sublata" (achievements) (1999a: 42) and Rob Kitchin's "capta" (taken):

> [...] what we understand as data are actually capta (derived from the Latin capere, meaning 'to take'); those units of data that have been selected and harvested from the sum of all potential data [...]. (2014b: 2)

As much as I acknowledge the critical consideration of data as 'given,' I would argue that the essence of data within scientific practice increasingly comes closer to its etymological origin. As Sabina Leonelli has shown, 'data' can only be considered as such when it is packaged for circulation. While I do not agree completely that this describes today's situation, the increasing mainstreaming of open data might soon make portability (or 'give-ability') conditional for 'data' to be considered as such in science.

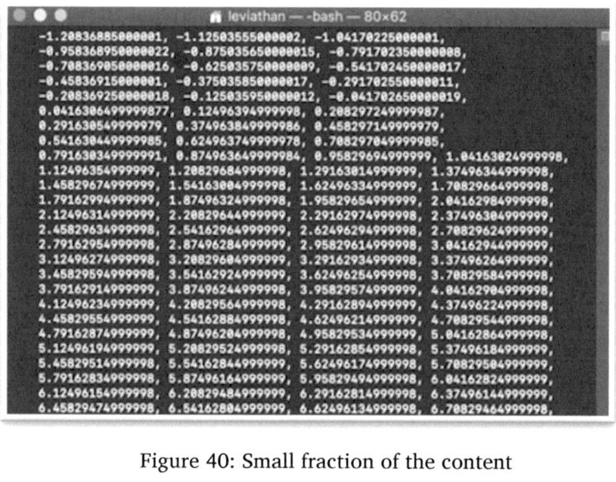

Figure 40: Small fraction of the content from file 'GCP_PPP-2005_1850-2100.nc' Source: My own screenshot

In my opinion, the term 'raw data' is actually less problematic than 'data' in general. At least in the case of computational science, it has a distinct meaning and justification in practice. 'Raw' refers to the data that is machine-readable but not manipulated to enable human cognition. In everyday language, scientists may point to columns of numbers when referring to 'raw data,' as shown in Figure 40.

The snapshot depicts a small fraction of the content from file 'GCP_PPP-2005_1850-2100.nc' as retrieved by the netCDF reader (‚ncdump –ct' command) and visually represented by my MacOS terminal. In a strict sense, the content represented through the snapshot is no longer 'raw data' but processed through algorithms to make it visible and cognizable for me as a human being. The raw version of the dataset cannot be shown as it is invisibly stored in a digital database[83] on my laptop's hard drive. Accordingly, when scientists talk about 'raw data,' they might point to visible columns of numbers, but they are understood as an index for the invisible information stored in the database at stake.

The use of the term 'raw data' is also rooted in principles of computational scientific practice. The scientists in my field always seek the maximum proximity to 'raw data' possible. Practically, this means that they prefer to work within text terminals or consoles, write their own software wherever possible and are generally skeptical toward software written by others, visual representations instead of numbers, and comprehensive proprietary tools. As a matter of fact, the frequent use of the term 'tool' in my interviews with climate impact modelers oscillate with Jörg Rheinberger's characterization of *technical objects* (1997) and its differentiation against *epistemic things*, which has also been discussed for the case of simulation models by Mikaela Sundberg (2008). In my

83 Marcus Burkhardt has shown that the term database is ambiguous in its meaning and use. It may refer equally to collections in general and collections of digital information in particular, i.e. the technologies that process structured collections of machine-readable information (2015: 131).

interviews, my informants referred to 'tools,' meaning software developed by others and beyond the control of the researcher:

> So, GRASS and Mapwindows, these are [...] tools that are used in this field. (Interview Willkomm)

> If you click here, [...] it's their tool that does everything. And they calculate everything internally, which then enables you to navigate here. (Interview Rechstein)

In contrast to epistemic things, such as 'question-generating machines' (e.g. the simulation models), tools are understood as technical objects or 'answering machines' (Rheinberger 1997; Sundberg 2008). They are problem solvers, not knowledge producers. As scientists are aware of the frictions between these two functions (Knorr-Cetina 2003), they traditionally refrain from using technical constellations with a (perceived) high epistemic opacity (more on these aspects in Chapter IV).

The differentiation between objects within and beyond the control of the scientist is also valid for the case of data, which bows down to the distinction between 'primary' and 'secondary data.' Primary data refers to data generated by the researchers themselves, making use of their own instruments, according to their proper research design and methodology (Kitchin 2014b: 7). By contrast, researchers often make use of *secondary data* generated and provided by others, possibly with very different instruments and research designs. In the case of the modeling work described here, for example, the researchers have used externally generated data to produce their own global dataset ('primary data') of historic and future GDP. This secondary data comes from heterogeneous resources and actors, such as the Madison Project

Database[84] (University of Groningen) and the Global Wealth Databook[85] (CS Credit Suisse bank). This fluidity of (i.e. 'open') data between modeling groups of different institutions, scientific fields (physics, economics) and production contexts (scientific, commercial) creates new opportunities but also challenges of trust, which are addressed by numerous strategies of standardization, documentation and evaluation.

The data files in our example, for instance, are encoded in widely standardized formats, such as NetCDF and comma-separated values (commonly known as a csv). The dataset is equipped with a digital object identifier (seen as DOI in bibliographies[86]), a format for unique resource identification. Kitchin highlights that such *indexical data* enable identification and linking;

> [...] indexical data are important because they enable large amounts of non-indexical data to be bound together and tracked through shared identifiers, and enable discrimination, combination, disaggregation and re-aggregation, searching and other forms of processing and analysis. [...] Indexical data are becoming increasingly common and granular, escalating the relationality of datasets. (Kitchin 2014b: 8)

The dataset also includes four metadata files in xml (eXtensible Markup Language) format, according to the standards iso19115, datasite, dif and escidoc. *Metadata* are essentially data about data.

> Metadata can either refer to the data content or the whole dataset. Metadata about the content includes the names and descriptions of specific fields (e.g., the column headers in a spreadsheet) and data definitions. These metadata help a user of a dataset to understand its composition and how it should be

84 https://www.rug.nl/ggdc/historicaldevelopment/maddison/releases/maddison-project-database-2018, retrieved on December 10, 2020.

85 https://www.credit-suisse.com/about-us/en/reports-research/global-wealth-report.html, retrieved on December 10, 2020.

86 http://doi.org/10.5880/pik.2017.011, retrieved on December 10, 2020.

> used and interpreted, and facilitates the conjoining of datasets, interoperability and discoverability, and to judge their provenance and lineage. (ibid.: 8f)

The formatting of the files in the markup language XML ensures that the information stored is both human- and machine-readable. Figure 41 depicts a snapshot of the iso19115-file[87] in its "raw" machine-readable version for illustrative purposes.


```
<gmd:referenceSystemInfo>
  <gmd:MD_ReferenceSystem>
    <gmd:referenceSystemIdentifier>
      <gmd:RS_Identifier>
        <gmd:code>
          <gco:CharacterString xmlns:gco="http://www.isotc211.org/2005/gco">ur
        </gmd:code>
      </gmd:RS_Identifier>
    </gmd:referenceSystemIdentifier>
  </gmd:MD_ReferenceSystem>
</gmd:referenceSystemInfo>
<gmd:identificationInfo>
  <gmd:MD_DataIdentification>
    <gmd:citation>
      <gmd:CI_Citation>
        <gmd:title>
          <gco:CharacterString xmlns:gco="http://www.isotc211.org/2005/gco">
            Spatially-explicit Gross Cell Product (GCP) time series: past obser
            (2010-2100)
          </gco:CharacterString>
        </gmd:title>
        <gmd:date>
          <gmd:CI_Date>
            <gmd:date>
              <gco:Date xmlns:gco="http://www.isotc211.org/2005/gco">2017-11-30
```


Figure 41: Snapshot of a metadata file according to iso19115.
Source: My own screenshot

In addition, a hyperlinked bibliography cites other datasets and scientific publications, which are related to the present set.

Linked data

There has been a shift in the understanding of open data from a category of human-human and human-machine interaction to a category of

87 https://dataservices.gfz-potsdam.de/pik/download.php?item=escidoc-2740907&mdrecord=iso19115.

machine-machine readability as I could witness in my fieldwork at the PIK and during an Open Science fellowship at the Wikimedia Foundation. As an example of the latter, we can quote from the FAIR data principles, which highlight the significance of "computational stakeholders":

> Humans, however, are not the only critical stakeholders in the milieu of scientific data. Similar problems are encountered by the applications and computational agents that we task to undertake data retrieval and analysis on our behalf. These 'computational stakeholders' are increasingly relevant, and demand as much, or more, attention as their importance grows. (Wilkinson et al. 2016: 2)

This coincides with a blending of the open data discourse with the idea of *linked data,* which is the idea to transform the internet from a 'web of documents' to a 'web of data':

> Such a vision recognises that all of the information shared on the Web contains a rich diversity of data – names, addresses, product details, facts, figures, and so on. However, these data are not necessarily formally identified as such, nor are they formally structured in such a way as to be easily harvested and used. (Kitchin 2014b: 52)

Tim Berners-Lee (2009) mentions four rules of behavior for such linked data in a semantic, machine-readable web: The first is to identify things with Unified Resource Identifiers (URIs): "If it doesn't use the universal URI set of symbols, we don't call it Semantic Web" (ibid.) The second rule is to use a particular type of such identifiers, namely Hypertext Transfer Protocol (HTTP) URIs, which works well with the inherent structure of the internet. The third rule is that one should provide meta-information on the web against a URI. Berners-Lee mentions a number of standards (RDF, XML) that can be searched by dedicated query languages (SPARQL). The fourth important rule is to produce links to these URIs elsewhere, "which is necessary to connect the data we have into a web, a serious, unbounded web in which one can find all kinds of things, just as on the hypertext web we have managed to build" (ibid.). A prominent example of a linked data project is the open knowledge base *Wikidata,* a machine-readable supplement and further development of the ideas around Wikipedia. In contrast to Wikipedia, the information (data) on Wikidata is highly structured in the ways described above.

As an example, we can take a look at the Wikidata entry for 'data'[88] (see table 3). Items are uniquely identified by a 'Q' followed by a number, in this case 'Q42848.' Statements describe detailed characteristics of an item and consist of a property (e.g. instance of) and a value (e.g. abstract object). Properties are identified by a 'P' followed by a number, such as 'subclass of (P279).'

Item	**data (Q42848)**
description	facts represented for handling
statements	
instance of (P31)	abstract object (Q7184903)
subclass of (P279)	information (Q11028)
part of (P361)	data base (Q59138835)
topic's main category (P910)	category: Data Q6641340
different from (P1889)	knowledge (Q9081)
identifiers	
JSTOR topic ID (P3827)	scientific data
(...)	

Table 2: Wikidata entry for 'data' (Q42848). Source: My own table

88 https://www.wikidata.org/wiki/Q42848, retrieved on December 10, 2020.

By means of Wikimedia's query service,[89] one can search through the Wikidata knowledge base and conduct comprehensive activities of data analysis.

In the context of our study, the Wikidata project may serve as an example for the sociotechnical imaginary (Jasanoff/Kim 2015) of a web of data. As a matter of fact, it is an imaginary not only supported by civil society organizations, such as Wikimedia, but also by powerful actors of the knowledge economy. From the beginning, Wikidata had been co-funded by Google and the Allen Institute for Artificial Intelligence, an organization established by Microsoft co-founder Paul Allen.[90] These actors are highly interested in the mainstreaming of linked open data, which promises numerous market opportunities. In this perspective, Google recently launched its own dataset research engine,[91] which enables one to search the web for structured datasets, and research data specifically. Within a web of data, the spheres of science and private business become ever more permeable. This permeability of data and its production contexts creates new challenges for scientific practice, as the following interview excerpt shows:

> We will derive a damage function, with data from the reinsurance company Munich Re, and also from SwissRe. And the data is not publicly available. And we use it to derive these damage functions, which can then be used by anyone. But in order to be able to reproduce the loss function, you have to get this data from Munich Re yourself. [...] There are a few other datasets that are publicly available for these damages, but they have some kind of spatial and also a temporary bias. (Interview Geiger, translated by the author)

89 https://query.wikidata.org/, retrieved on April 3, 2019.

90 http://tcrn.ch/H0aO9U, retrieved on April 3, 2019.

91 https://toolbox.google.com/datasetsearch, retrieved on April 3, 2019

The impossibility of reproducing the scientific process due to the use of proprietary information as input data for simulations is problematic from a perspective of scientific verifiability and political accountability. Moreover, it creates a shift in the distribution of work between scientific and other actors, which can also be a trigger for controversy:

> Yes, that is in a way an outsourcing of their own research activities. [...] So, there are different opinions on that. The Munich Re clearly says that they are very interested in doing research with it. Because they benefit from it. Swiss Re then seems to be a bit more reserved [...]. In addition, there are others, pseudo-insurers, or over-insurers in the United States, who simply earn a lot of money with these data, by selling them to the insurance industry. They don't make them freely available. (Interview Geiger)

Of course, the protection of scientific independency from economic interests is not a new issue, but it gains new relevance in the context of 'climate services' (Krauss/von Storch 2012; Vaughan/Dessai 2014), and data-driven and sustainability research in general.

Open data infrastructure

As we have already seen in some of the examples discussed above, open data is not just a matter of appropriate licensing but requires the setup of comprehensive new infrastructures. In Chapter I, I briefly mentioned the repurposing of the existing infrastructural base at the Science Park Albert Einstein for contemporary techno-scientific challenges. The most virulent example is the repurposing of the previous Geodetic Observatory into the library shared by all institutes of the science park. The main reading room of the library is located at the former 'great instrument hall.' The 'hall of the pendulum,' which had hosted the consequential geodetic experiments by Friedrich Jakob Kühnen and Philipp Furtwängler (Kühnen/Furtwängler 1906; Reicheneder 1959), has now been converted into a museum of geodetic instruments (see Figure 42).

Figure 42: Museum of geodetic instruments at the library of Telegrafenberg. Source: GFZ (2012)

In fact, the library is distributed in several buildings on the hill, including the PIK headquarters at Michelson House. This productive collaboration between the four institutes on the hill is not self-evident, given the extraordinary ambience of competition between the organizations. As a former employee of the GFZ IT services highlights:

> This becomes increasingly relevant, considering that the library also serves as data repository for so-called 'gray literature,' which includes software and data.
> (Interview Gephardt, translated by the author)

Co-funded by the institutes on the hill, the library is currently investing many resources to build such data infrastructure. This includes the hosting and documentation of datasets such as the one discussed above and a sophisticated system and interface to make the data accessible. One can browse a global Leaflet map (see Chapter VI), for example, to search for datasets that address a specific region. Figure 43 shows such a spatial representation of data available for the PIK's dataset 'Simulation Data from Water (regional) Sector.'[92]

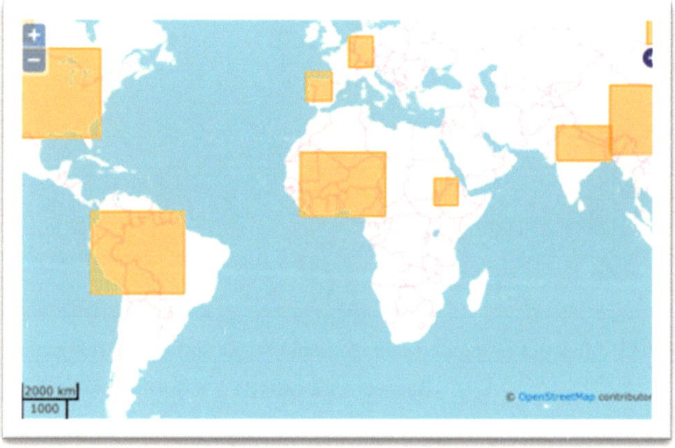

Figure 43: Leaflet map of the spatialities of a dataset.
Source: GFZ Data Services

We can see that 'open data' is not only a matter of saving datasets in an appropriate format or adding meta-data to it, but also involves the development of a comprehensive new infrastructure, with its own entanglement of physical, symbolic and – not to forget – human elements.

Open data is work

This last aspect – human engagement – is often forgotten in the discussion about open data. In my interview with IT infrastructure expert Paul Gephardt, he highlights that open data infrastructures will have considerable consequences for the life realities of the scientists:

92 http://dataservices.gfz-potsdam.de/pik/showshort.php?id=escidoc:2959917, retrieved on April 3, 2019.

> [...] research data management plans will be necessary, the provision of safe data repositories, data must be labelled with Digital Object Identifiers. And to implement the entire transformation process towards an open science paradigm into reality.
> (Interview Gephardt)

It seems useful to reconsider open data as an objectual category and investigate more thoroughly what 'opening data' means as a practice. Michael Gurstein, among others, has criticized this sole focus on objectual characteristics of open data:

> As an object or thing the attributes and characteristics of the open data are more or less fixed once made available to the end user/consumer. As well, the determination of the attributes or characteristics of the data (what the open data "is") as seen/obtained by the end user is solely at the discretion of the producer and are uniform and stable as between end users. (2013)

Instead of such an understanding of open data as objects and products, Gurstein proposes a focus on open data as a service:

> But why shouldn't we think of 'open data' as a 'service' where the open data rather than being characterized by its 'thingness' or its unchangeable quality as a 'product', can be understood as an on-going interactive and iterative process of co-creation between the data supplier and the end-user [...]. (ibid.)

In so doing, one could put more emphasis on 'opening' as a transitive and interactive process, an interaction and relationship between suppliers and users. For Gurstein, this reconceptualization would have consequences for the way in which open data is funded, managed and made available. It would require a review of the relationship between the open data discourse and neoliberal agendas marketizing public services.

> [...] if one treats open data simply or exclusively as a thing or commodity then it is available solely as a product for purchase and use through the market place–where of course, market principles dominate and where for example, those with the most resources are able to command and control and thus precipitate the supply of the product i.e. the open data. (ibid.).

This oscillates with the 'hijacking' of the open data discourse by the big players of the knowledge economy discussed above. In an earlier article written in 2011, Gurstein proposes a number of necessary elements enabling a more effective and inclusive use of open data, thereby reducing the "data divide." These include:

- available telecommunications/Internet access;
- having access to machines/computers/software;
- having sufficient knowledge/skill to use the software required for the analyses;
- having the data available in a format to allow for effective use at a variety of levels of linguistic and computer literacy;
- sufficient knowledge and skill to see what data uses make sense (and which do not) and to add local value;
- having supportive individual or community resources sufficient for translating data into activities for local benefit; and
- the required financing, legal, regulatory or policy regime, required to enable the use to which the data would be put.

(summarized from Gurstein 2011: 5f)

Within the years, general agreement on these challenges and needs have enabled large resources to build up open data infrastructures and services. Examples are the Data Services at the Science Park Albert

Einstein discussed, and much more extensive initiatives, such as the National Oceanographic and Atmospheric Agency (NOAA) Data Discovery Portal[93] in the United States or Worldbank Open Data.[94] The strategic importance of such infrastructures and data centers became clear at the time of the inauguration of US President Donald Trump. Due to his regular statements describing climate change as a hoax, climate scientists and civil society organizations feared that the new president would shut down the research programs on climate change, leading to a deletion of climate-related data stored on government servers, for example, data hosted by the NOAA, the Environmental Protection Agency and the White House. Based on such fears of data demolition, scientists from Penn and other Universities organized DataRefuge, a large-scale data migration of US government data to servers in Canada. While the scenario of data demolition has not yet materialized,[95] the episode shows how open data is dependent on working infrastructures and institutional support. As a matter of fact, the main value of the DataRefuge project might not have been to copy datasets from one server to the other. Rather, the project triggered activities of infrastructural inversion: While environmental data has formerly been taken for granted, the wide-ranging discussion around DataRefuge generated a variety of initiatives surfacing datasets, considering their relevance and characteristics, and making them more effectively available. The chief data officer of NOAA, Edward J. Kearns, for example, published a comprehensive statement, where he mapped out the whole data infrastructure of the institution and addressed the public worries considering politically motivated data deletion:

93 https://data.noaa.gov/datasetsearch/, retrieved on April 3, 2019.

94 https://data.worldbank.org/, retrieved on April 3, 2019.

95 https://sunlightfoundation.com/tracking-u-s-government-data-removed-from-the-internet-during-the-trump-administration/, retrieved on April 2, 2019.

> I am sometimes asked if NOAA's data in its archives can be easily deleted. No they can't, since data may not be removed without significant effort and public deliberation. It is also unlawful to tamper, damage, delete, vandalize, or in any way alter formal federal records, including NOAA's environmental data and its archives.[96]

Figure 44: DataRefuge logo. Source: www.datarefuge.org

Kearns also criticized an exaggerated focus on the datasets as objects, ignoring the crucial contribution of open data work within research infrastructures:

> The value of NOAA's data archives include not just the simple existence of the data themselves, but the continuous investment of NOAA's experts' efforts towards the sustained quality and usability of the data. The integrity and accuracy of data that are stored on non-federal system and are not stewarded by NOAA's scientists cannot always be easily verified beyond file-level distribution. NOAA is currently exploring best practices and technologies that may allow the authentication of its data throughout the wider data ecosystem, and welcomes interested parties in academia and industry to join in this exploration.[97]

Notwithstanding these insurances by Kearns, the activists of DataRefuge continued to move datasets from governmental servers, which turned out to be extensive and tricky work. The activists had to invent new methods of identifying, understanding, copying, ordering, monitoring and making datasets accessible. Given the effort and inventive-

96 https://libraries.network/blog/2017/4/30/on-the-preservation-of-and-access-to-noaas-open-data, retrieved on April 2, 2019.

97 https://libraries.network/blog/2017/4/30/on-the-preservation-of-and-access-to-noaas-open-data, retrieved on April 2, 2019.

ness, they came up with a new job description for these tasks: The 'data baggers' write custom scripts to scrape complicated datasets from distributed sources and patched-together federal websites. A coverage of the *Wired* magazine shows that DataRefuge led to a veritable imagination of a parallel infrastructure monitoring research infrastructure:

> [...] two dozen or so of the most advanced software builders gathered around whiteboards, sketching out tools they'll need. They worked out filters to separate mundane updates from major shake-ups, and explored blockchain-like systems to build auditable ledgers of alterations. Basically it's an issue of what engineers call version control – how do you know if something has changed? How do you know if you have the latest? How do you keep track of the old stuff? [...] DataRefuge and EDGI understand that they need to be monitoring those changes and deletions. That's more work than a human could do. So they're building software that can do it automatically.[98]

Open research data are not as fluid as one could think, but rather a sticky matter. As the example of DataRefuge shows, seamless data portability imagined by the open data discourse is a tricky object of desire. The more one wants to mobilize datasets and reduce the seamfulness in systems (Vertesi 2014), one is obliged to invest in the construction of a fluidifying data infrastructure.

Infra-worlds of knowledge

Based on the preceding discussions, I would like to come back to the literature discussed at the beginning of this chapter and propose some reconsiderations of the status of research data in contemporary computation science. As Rheinberger highlights, DNA sequencing has been delegated to automated analyzers for several decades now. The field of bioinformatics has been developed with its own set of methods and

98 https://www.wired.com/2017/02/diehard-coders-just-saved-nasas-earth-science-data/, retrieved on April 2, 2019.

infrastructures to domesticate this plethora of data. For Rheinberger, this represents a new phase in the relationship between (molecular life) science and information. While it was formerly the discursive and conceptual aspects of information that have been prominent in molecular genetics, bioinformatics has shifted this focus to the sphere of the infrastructural:

> Data have become a resource, rather than a result in the world of infra-experimentality, produced on an industrial scale and made intelligible only in the context of appropriate software. The research technologies in the space between the knower and the to-be-known have entered the stage of a second order mediation. The data, mediators between traces and technophenomena, have proliferated and created a world of their own. (Rheinberger 2011: 346)

These "infra-worlds of knowledge" (ibid.) in the field of simulation modeling do not only gain momentum in daily scientific practice but also as a matter of scientific reputation. *Inter alia,* this valuation is demonstrated by the rise of the *data publication.* While there is significant debate around the formats, processes and terminology of this new publication format, the general purpose is to "bring datasets into the scholarly record as first class research products (validated, preserved, cited, and credited)" (Kratz/Strasser 2014: 1). The format promises deliverables for various actors, including scientists, journal editors, publishers, data centers, the scientific community, funding agencies, governments and society as a whole. According to a conceptual paper from 2009 (Costello 2009: 420), benefits include additional publications and higher citation rates for individual researchers, possibilities for verification and accountability, greater valuation of data and data producers, higher financial return on research investments and, simply put, "better science."

We can illustrate the concrete realization of the format in scientific practice with the example from the PIK relating to the dataset discussed above: *Continuous national gross domestic product (GDP) time series for*

195 countries: Past observations (1850–2005) harmonized with future projections according to the Shared Socio-economic Pathways (2006–2100) (Geiger 2018). The data publication appears in the journal *Earth System Science Data*, which presents itself as "an international, interdisciplinary journal for the publication of articles on original research data (sets), furthering the reuse of high-quality data of benefit to Earth system sciences."[99] The publication begins with a discussion of the underlying metric for the dataset, the GDP. The author describes the GDP's role as a standard indicator for assessing a nation's development and discusses the criticisms regarding its representational features regarding growth, development, and welfare and well-being. Despite these limitations and because of a lack of alternatives, "GDP has proven to be a useful measure to track the evolution of economic development within or across nations" (ibid.: 487).

The publication then describes input data and the methods used to create the dataset in question: A continuous and consistent GDP time series for 195 countries. Input data includes the Penn World Table, the Maddison Project Database, World Development Indicators, the History Database of the Global Environment and future projections from the SSPs. The discussion in the methodological section addresses ways to deal with missing data and interpolation:

> As a first step we populate all missing data points in 1850, the initial year of our data product, by linear interpolation between the last available data point before 1850 and the first one after 1850, ensuring that it is not more distant in time than 1870. Next, and if available, we generate annual data by linear interpolation of data points between 1850, 1860, and 1870. These preparatory steps reduce the missing value fraction from 51.7 to 48.5 %. (ibid.: 850)

99 https://earth-system-science-data.net/, retrieved on April 2, 2019.

The text focuses specifically on calculation issues for particular regions of the world. The Balkans, for example, create considerable challenges for calculability due to frequent territorial alterations and missing data in times of conflict. Equally, data quality for the African continent is extraordinarily low:

> The MPD [Maddison Project Database] contains only six countries with income data prior to 1950: Egypt, Tunisia, Morocco, Algeria, and South Africa/Cape Colony (all since 1820), and Ghana (since 1870). Therefore, the African total (AT) population-weighted average income prior to 1950 is only defined by six countries. For historic and geographic reasons, we assume that those countries define the upper income limit when extrapolating the remaining countries back in time. (ibid.)

From a representational point of view, data for this region is generated basically by the author alone and only rarely represents original data produced 'on the ground.' After the explanations of the method and process, the publication describes the resulting datasets: "We provide three different primary data sets, a data description file, and two supplementary data sets in the online archive at https://doi.org/10.5880/pik.2017.003" (ibid.: 854). In the last section of the paper, the author makes an assessment of the quality of the dataset:

> While rather exhaustive data exist for Western European countries, these limitations might be less of a problem than for most African countries. As a consequence, one should treat the data with care and allow for uncertainties, in particular where data coverage is limited or almost non-existent. (ibid.)

"Treat the data with care" refers to the prospective users of the dataset. This reuse by scientific or nonscientific actors is the primary objective of the data publication (understood as the practice and resulting set of artifacts). However, as has been mentioned earlier, it also becomes a matter of reputation to make datasets public. Thanks to the format of the data publication, datasets become citable within global citation databases, such as the Web of Science,[100] Scopus and Google Scholar. These bibliometric and scientometric evaluation schemes of science (Sengupta, 2009) increasingly dominate the reputation systems of the researchers and their institutes. The relevance of these schemes can be illustrated by a tweet of the PIK researcher Stefan Rahmstorf, representing his status in the "top 1 % of the world's most-cited researchers in the field of Geosciences" on the Web of Science (see Figure 45). The tweet[101] has been pinned to Rahmstorf's feed for almost a year.

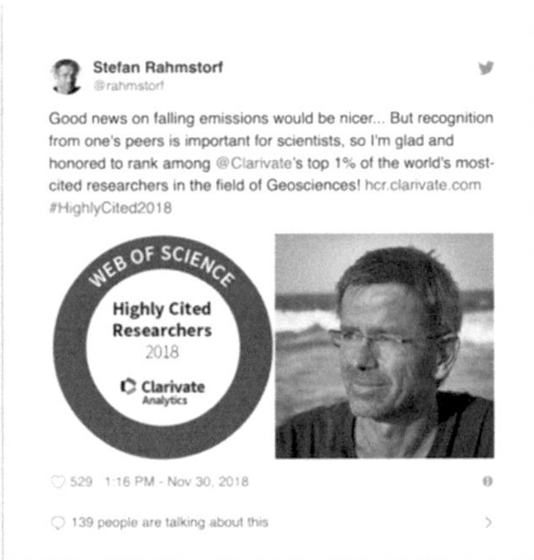

Figure 45: Pinned tweet by PIK researcher Stefan Rahmstorf. Source: Twitter

Against this background, data publication becomes increasingly important as a device to perform in these reputation schemes. By producing and publishing data with an open license requiring attribution (e.g. CC BY), one makes sure that all future computational work involving the datasets generates a citation in bibliometric schemes.

100 https://clarivate.com/webofsciencegroup/solutions/web-of-science/, retrieved on April 2, 2019.

101 Tweet posted on November 30, 2019, retrieved on September 3, 2019, via https://twitter.com/rahmstorf/status/1068463676628312064.

I would argue that this has significant consequences for the scientific practices in computational science. As the example of the GDP dataset shows, the aim of published data is not to create a new epistemic thing but a technical object, in the sense described by Jörg Rheinberger (1997). It is a forward-looking scientific practice aiming at the production of a robust base for further calculations. Literally and metaphorically, the dataset constitutes a spatiotemporal 'grid' that will drive further simulations. Against this background, the choice to base this grid on GDP is only logical. It makes sure that the dataset is not only used within the world of climate research but can potentially be used by virtually anyone simulating economic and social development in the world. The compatibility of the time series with future scenarios of socioeconomic development generated at the PIK (i.e. SSPs) makes it possible to use the dataset as a base for all sorts of predictions for the 21st century. More than 'facts,' 'information' or 'evidence,' these datasets constitute infrastructural elements for all sorts of prospective knowledge claims to be generated by scientific and nonscientific actors. Against this background, I would refer to them as *infrastructural data,* as they can build a fundament for multiple future work.

In my interviews, climate modelers often used the terms of 'drivers' and 'drivers of the future.' 'Drivers' refer to computer models that provide the conditions for other models. Such drivers are models that drive other models further down the model chain.

> This means that the driving model is always a global model, and the regional model then provides much more precise climate information for Greater Europe, due to the fact that the topography is resolved much more finely. (Interview Hauser, translated by the author)

On a material level, it is not the model that drives future calculations but its outputs in the form of datasets, for example, a time-series of GDP, temperature or CO_2data. The focus of work in the field of climate impact research is increasingly shifting to data(sets). While climate modelers tend to dislike the characterization of the 'data scientist,' their field is surely subject to datafication on various levels. We might speak of a datafication of computational science in this context, rather than of data-science. Nevertheless, the Potsdam Institute recently used the connotation of the 'data scientist' for the first time in a job advertisement. The PIK is also a member institution of *GeoX Data Science*,[102] an innovation program funded by the Telegrafenberg institutions and seven other geoscientific facilities in the Berlin-Brandenburg area.

A pragmatist typology of data

We should reconsider the status of categories such as 'data,' 'research data' and 'evidence' carefully. Being conscious of the limits of comparison, we can draw some parallels between Rheinberger's observations in the field of bioinformatics and climate-related simulation modeling. Regarding the example of the *Spatially-explicit Gross Cell Product (GCP) time series* discussed above, we have seen that the relevant dataset includes a variety of different 'data,' which may individually be labeled as primary data, secondary data, tertiary data, indexical data, metadata, and so on. The label does not characterize any essential (material) qualities of the data but its relational entanglement and situated context of use. As Leonelli highlights: "Depending on what uses the data are eventually put to, and by whom, those modifications may well prove as relevant to making data into valuable evidence as the efforts of the original data producer" (2015: 8). While I agree with this argument, we may want to reconsider the predominant role that Leonelli gives to the category of scientific data as 'prospective evidence.' Instead, I propose a categorization of data according to its concrete use in research practice, which goes well beyond its prospective use as scientific evidence.

102 www.geo-x.net/en/, retrieved on April 3, 2019.

Pragmatist categories of research data:

1. Evidential data

Evidential data serve as proof of a certain attribute of a phenomenon and establish scientific facts. In climate impact research, evidential data may represent predictions regarding future developments, evaluations of future risk or correlations between such variables.

2. Infrastructural data

Infrastructural data are not produced with the aim of providing evidence but to provide a stable foundation for manipulations of other data. They are evaluated, refined and optimized to serve these purposes. The becoming of infrastructural data requires cumbersome work of testing, standardization and evaluation. This means that the becoming of infrastructural data takes time and involves cooperative efforts in communities. Examples in the context of climate simulation modeling are standardized geospatial databases and time series (e.g. the series of GDP values discussed earlier).

3. Resourceful data

In this context, data serves as a resource for the identification of patterns, thereby enabling the extraction of further information and knowledge. It is the role of data within practices referred to as 'data-driven research' or 'data-science.' A popular example of resourceful data is ImageNet,[103] a large image database currently containing more than 14 million images, classified along object properties. ImageNet has

---

103 Description from the ImageNet website: "ImageNet is an image database organized according to the WordNet hierarchy (currently only the nouns), in which each node of the hierarchy is depicted by hundreds and thousands of images. Currently we have an average of over five hundred images per node. We hope ImageNet will become a useful resource for researchers, educators, students and all of you who share our passion for pictures." Retrieved on September 23, 2019, via http://www.image-net.org/.

been used most prominently as a training dataset for convolutional neural networks since the 2012. Another example is the use of social media data in the context of *digital methods* research (Rogers 2013; 2015) and its dedicated software tools (e.g. Borra/Rieder 2014; Rieder 2015). As a matter of fact, researchers involved in the Digital Methods Initiative have often used a 'climate change dataset' within their studies, given its quality as a dataset. We can exemplify this with a text passage from Noortje Marres and Carolin Gerlitz's article on *Interface Methods:*

> For our analysis of 'happening content,' we decide to focus on a fairly general issue term, namely *climate change*, and include in our data set all Tweets using this term for a period of almost three months – from March 1st, 2012 to June 15, 2012, adding up to a total of 204795 tweets, a workable, medium-sized data set. (2015: 17)

In this context, 'climate change data on Twitter' was not primarily chosen in order to make statements about public debates on climate change but due to its quality as a "workable, medium-sized data set."

4. Communicational data

Communicational data enable the mobilization and resulting portability of evidential, infrastructural or resourceful data. In so doing, communicational data permits cooperation and prospective reuse by agents such as human researchers, machines and hybrid collectives. Examples of communicational data are meta-data, linkages and identifiers, which have been discussed in this chapter previously.

Taken together, the four categories are the essential ingredients populating the "infra-worlds of knowledge" (Rheinberger 2011: 346) of fields such as computational and data-science. Given the relationality and interpretative flexibility of data, these categories are not exhaustive and their boundaries are not solid. They are not substantive but relational and dependent on their situational context. 'Datasets' (relational assemblages of data) in the age of open science are typically structured in a way to enable their manipulation in several categories.

Our GDP dataset, for example, is evidence for a probable spatiotemporal distribution of GDP values. It is also infrastructural data for prospective calculations and involves mobilizing data, such as meta-data, linkages and identifiers.

To assess the belonging and aptness of data in those categories is one of the essential skills that differentiate professional 'data-scientists' from other actors working with digital data (e.g. computational scientists, computer scientists, statisticians). This skill will be of increasing importance in a scientific environment subject to comprehensive 'datafication' to ensure effective and sound scientific practices. It will also require new technological devices that can help to assess data quality and make its characteristics accountable.[104]

104 An example of such a device is OpenRefine, a "tool for working with messy data: cleaning it; transforming it from one format into another; and extending it with web services and external data." Self-description from http://openrefine.org/, retrieved on April 2, 2019.

VI. Future fluidity

At the beginning of the research for this study, I traveled to San Francisco to hold a presentation at the annual meeting of the American Association of Geographers. Being there also provided an occasion to conduct many interviews with American climate and water risk modelers. These researchers were at the forefront of the transformations of scientific practice, which have been discussed within this study. Accordingly, the journey to the United States provided me with a taste for developments that would materialize on Telegrafenberg a few years later. As a matter of fact, the Telegrafenberg institutes are just now raising similar infrastructures as those discovered in my journey to California in 2016. Similar to the time-spaces of our fields, our own research entails complex nested temporalities, including those of the past, present and future.

Engaging in data amplification

This chapter aims at operationalizing some of the arguments and investigative devices developed so far. It does so discussing the case of *Surging Seas,* a comprehensive digital mapping of flood and SLR enabled through common efforts by a multitude of actors. I would argue that these activities can provide a glimpse of what predictive work driven by open data and digital platforms could look like in the future. While Surging Seas is not a scientific project in its own right, the example also shows how permeable the worlds of science, politics, the private sector and 'the public' have become within the sphere of the digital. The artifacts populating such digital constellations are fluid. Consequently, substantial characterizations of such objects will be ephemeral and of limited value. It is the relationality and its evolvement that matters in these constellations.

As Rob Kitchin has shown, a variety of actors are putting considerable effort into the establishment of relationships between different sets and kinds of data. Drawing on a note by Jeremy Crampton, Kitchin refers to this process as *data amplification:*

> [...] that is, data when combined enables far greater insights by revealing associations, relationships and patterns which remain hidden if the data remain isolated. As a consequence, the secondary and tertiary data market is a multi-billion dollar industry [...]. (2014b: 8)[105]

What does such data amplification look like in practice? And how can we make sense of the way actors are making sense of data? How can we represent relationships of 'amplified data'? This may well bow down to a question of diagrammatics. What is the form of our own mental representations of knowledge? What is the diagrammatic form we are comfortable to work with? Is it a chain of references, as in Latour's cascade of inscriptions? Or a rhizomatic network of relationships, consisting of nodes and edges? Or a triangle, representing the layers of a knowledge pyramid? The scientists in my field are confronted by the same challenges when they structure and visualize their digital datasets. The way they deal with it could be described as *kneading.* Within this practice, data is repeatedly arranged in various forms using algorithmic inscription techniques. The aim is not to find the best way to 'represent phenomena as realistically as possible,' but to experiment with the materiality of the data – its texture, robustness and resisting power. The Jupyter Notebook entangled with the Python visualization package Matplotlib (see chapter IV) is a good example of an

105 For the term data amplification, Kitchin refers to Crampton et al. (2013). However, the term actually does not appear in that publication.

environment enabling efficient and effective kneading. Here, data can be knitted (i.e. plotted) into various visual forms in seconds, and the resulting inscriptions may even be compared in parallel. In contrast to more comprehensive proprietary tools for data visualization (e.g. tableau.com), the fairly simple diagrams in Matplotlib do not give away much control over the plotting method.

I found this technique valuable for my own organization of data, kneading it repeatedly into different geometric forms. Kneading qualitative data is much more time-consuming than plotting structured datasets with Matplotlib, but it helped to gain different perspectives on the data captured from my field(s). One possible geometric representation of data and relationships is the knowledge pyramid, as represented by Rob Kitchin (2014b: 10).

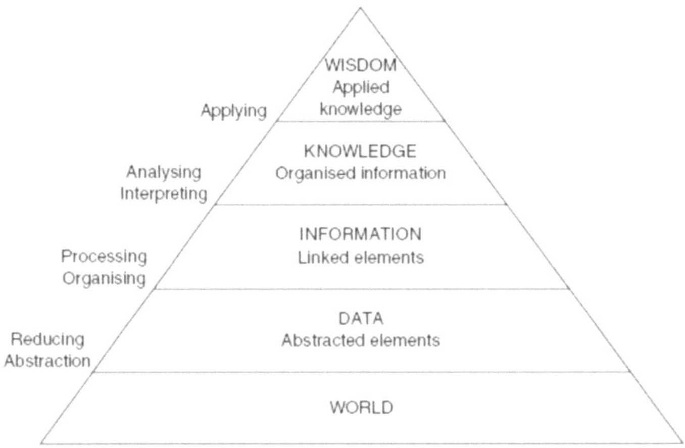

Figure 46: Knowledge pyramid. Source: Kitchin (2014b: 10)

Kitchin thinks of data as forming the base of a knowledge pyramid (see Fig. 46): “[D]ata precedes information, which precedes knowledge, which precedes understanding and wisdom” (ibid.: 9). The layers of the pyramid are distinguished by a process of distillation that “adds organization, meaning and value by revealing relationships and truths about the world" (ibid.). Stepping up the layers of the pyramid is enabled by practices of "reducing, abstracting, processing, organizing, analyzing, interpreting, applying" (ibid.). Kitchin's pyramid builds on similar visual representations of knowledge by Russel Ackoff (1989), Mortimer Adler (1986) and David Weinberger (2014). The analytical consistency of the pyramid has been critically evaluated and partially contested by various authors (Frické 2009; Rowley 2007). Whether the layering of the pyramid appears meaningful depends on the way we understand its constituting elements[106] However, as mentioned earlier, such diagrams may be understood more as tools for kneading our understanding of captured data rather than as representations of the world. Specific geometric forms might also fit particular fields and be estranged from others. For me, it represents well how the actors in my field generally understand the status of their data, their work with information and their creation of climate knowledge, hopefully enabling better ways of handling the present and the future ('wisdom'). However, the knowledge pyramid does not tell us much about the way data, information, knowledge and wisdom may actually be assembled in practice.

During the research and my interviews in California, I discovered two interrelated constellations which are formative for the way climate change-related SLR is imagined and debated in the United States: NOAA's Digital Coast, and Climate Central's Surging Seas. As I learned later, both initiatives are also connected to the PIK's work on global and local SLR. The two interconnected projects (Digital Coast and Surging Seas) provide an opportunity to investigate how actors engage in data amplification.

106 This depends greatly on one's school of thought, as can be illustrated by the multiple understandings of 'information' (Floridi 2010; Kitchin 2014b).

A Digital Coast

The NOAA is an American scientific agency within the United States Department of Commerce that focuses on the conditions of the oceans, major waterways and the atmosphere.[107] The NOAA and a consortium of partners initiated an immense project in 2010 aiming at the production, gathering and provision of high-resolution elevation data for the entirety of the US coastline. Concretely, this required numerous flights over the coastlines with airplanes carrying Light Detection and Ranging equipment (LiDAR): LIDAR, which stands for *Light Detection and Ranging*, is a remote sensing method that uses light in the form of a pulsed laser to measure ranges (variable distances) to the Earth. These light pulses – combined with other data recorded by the airborne system – generate precise, three-dimensional information about the shape of the Earth and its surface characteristics.

As the NOAA states on its website, "LIDAR systems allow scientists and mapping professionals to examine both natural and manmade environments with accuracy, precision, and flexibility."[108] The technology has been used for a variety of purposes and application fields, including archeology, cartography, policing and, recently, as a key technology enabling self-driving cars.

As illustrated by a NOAA report, LiDAR data has permitted massive improvements of the National Elevation Dataset and enabled applications in the field of coastal management:

> Lidar data represent several important improvements over previous and commonly used vertical data sets generated for U.S. Geological Survey (USGS) topographic quad maps. The data available through the National Elevation Dataset (NED) have largely been created using photogrammetric

107 See https://www.noaa.gov/ and https://en.wikipedia.org/wiki/National_Oceanic_and_Atmospheric_Administration, retrieved on April 2, 2019.

108 Summarized from oceanservice.noaa.gov/facts/lidar.html, retrieved on April 2, 2019.

> techniques. The resulting accuracy of the NED is on the order of 3 meters or 10 feet […] with 10- to 30-meter resolution. […] much of the NED are fairly old, have vertical accuracies that limit coastal applications, and have horizontal resolutions that preclude the definition of coastal features. Lidar, while similar in cost to photogrammetry, is a more rapid technique that relies largely on new technology to produce results. Note that the NED is being updated with lidar data as they become available, particularly for the newer 1/9th arc-second (about 3 meters) resolution NED, […]. (NOAA 2012: 13)

It is important to highlight that LiDAR data for individual locations may be much more precise than the three meters resolution stated above, but the data has to be averaged into a homogenous national dataset, requiring a downgrade resolution at the local level. One has to differentiate between the accuracy of LiDAR as a sensing technology and the resolution of final elevation models, resulting from cumbersome interpolation, integration and homogenization work. The outcome of this harmonization work is cast into the *Digital Coast,* a comprehensive open data infrastructure accessible through a web platform.[109] The example of the Digital Coast shows impressively that open data is about much more than just 'putting data online.' Drawing on Rob Kitchin and Tracey Lauriault's work, we can characterize the Digital Coast as a *data assemblage*:

> A data assemblage consists of more than the data system/infrastructure itself, such as a big data system, an open data repository, or a data archive, to include all of the technological, political, social and economic apparatuses that frames their nature, operation and work. (2014: 6)

The Digital Coast is about bringing countless primary datasets into a standardized format, making them identifiable and portable through a relational structure of additional information, providing visual tools for the analysis and further manipulation of the data, and offering 'open

109 coast.noaa.gov/digitalcoast/, retrieved on April 2, 2019.

data services,' such as training, and workshops for a variety of stakeholders. As such, Digital Coast has been style-building and an inspiration for many other elaborations of open data infrastructures, including the one on Telegrafenberg (see Chapter V). One has to be conscious of the fact that the Digital Coast is a multimillion dollar enterprise, which is only possible through and depending on comprehensive governmental and legislative support. This can be illustrated by the *Digital Coast Act S.110* of 2017, as reported on the website of the US Congressional Budget Office (CBO):

> S. 110 would authorize the appropriation of $4 million a year over the 2018-2022 period to continue the National Oceanic and Atmospheric Administration's (NOAA) Digital Coast program. Under that program, NOAA makes geospatial data, decision-support tools, and best practices regarding the management of coastal areas available on a public website. (In 2016, NOAA used $4 million of appropriated funds to carry out the Digital Coast program.) CBO estimates that implementing the bill would cost $20 million over the 2018-2022 period, assuming appropriation of the authorized amounts.[110]

The data assemblage of the Digital Coast has been an enabler for many other ground-breaking initiatives, including comprehensive imaginations of the future with climate change. At the same time, it also pre-structures the way such imaginaries look and behave, as becomes clear with the example of the SLR mappings by Climate Central. While the NOAA is providing datasets, tools and services to expert communities and stakeholders, the mission of the NGO Climate Central is to cast these data into different visualities and to mobilize and situate them within various public debates. As NOAA employees highlighted in an interview carried out in California:

110 https://www.cbo.gov/publication/52496, retrieved on April 7, 2019.

> So, we are working with these folks who are using this to make decisions about permitting and planning, but also to reach to local folks on the ground; stakeholders that are in cities and communities, to help them see the impacts to their infrastructure and communities. In contrast to Climate Centrals Surging Seas. So, that's for a much broader audience. That is to tell stories for journalists, for the general public. So it's very different.
> (Interview with two NOAA functionaries in San Francisco)

Surging Seas

Surging Seas is a thematic program on SLR and flood risks developed by the non-governmental organization Climate Central. According to its mission statement,

> Climate Central surveys and conducts scientific research on climate change and informs the public of key findings. Our scientists publish and our journalists report on climate science, energy, sea level rise, wildfires, drought, and related topics.[111]

The organization is headquartered at Princeton University and well connected to research facilities in the US and abroad. Collaborators include climate scientist Anders Levermann, who heads the PIK's research department on complexity science[112] and is Professor of Physics at Potsdam University. The scientific intersections between Climate Central and the PIK particularly include the issue of SLR and its projection into the future. In this context, Levermann and Climate Central's Benjamin Strauss and Scott Kulp have developed a study together calculating the number of people in the US threatened by long-term SLR:

111 https://www.climatecentral.org/what-we-do#wwd, retrieved on May 3, 2019.

112 Formerly 'transdisciplinary concepts and methods,' the host department for my field research.

> Based on detailed topographic and population data, local high tide lines, and regional long-term sea-level commitment for different carbon emissions and ice sheet stability scenarios, we compute the current population living on endangered land at municipal, state, and national levels within the United States. (Strauss et al. 2015: 1)

The scientific article includes a political message urging for severe climate protection measures:

> Although past anthropogenic emissions already have caused sea-level commitment that will force coastal cities to adapt, future emissions will determine which areas we can continue to occupy or may have to abandon. (ibid.)

While Climate Central has published a number of such academic articles, their focus lies clearly on the interface between science and non-scientific publics. The most prominent example is Surging Seas, a program arranged around the production of interactive flood maps and their embedding into various platforms of public debate.

Fluid maps

The work of Climate Central and its Surging Seas program crystallizes in compelling digital mappings of floods and SLR. If the NOAA's Digital Coast is to be characterized as data assemblage, we might want to employ the term 'mapping assemblage' for the case of Surging Seas. However, as certainly as they qualify as maps, they are far from a common understanding of maps as truth documents, representing the real world with a certain degree of precision (Dodge et al. 2011: 4). Critical cartographers have always challenged this view of maps as representations of the world, and their arguments oscillate significantly with the way maps work in Surging Seas. As we have seen previously, the cartographies in question are subject to considerable interpretative flexibility. This flexibility within the maps is wanted by its developers, and it is a major source of their power.

Mapping choices

Figure 47 depicts a snapshot of one of these cartographic enterprises. 'Mapping Choices' provides a visual comparison of two SLR scenarios for the same geographic area, here focusing on New York City. Areas at risk of SLR are tainted in blue, representing the surging seas. The story of the mapping is straightforward: Carrying on with business as usual, we have to account for many more risks and expect more damages than if we limit global warming to two degrees, the threshold agreed at the Paris climate summit in 2015. On a rhetorical level, it mediates a choice: While some of the impacts of climate change are preventable, we can lessen the most severe consequences by acting now. Politically, it situates the flood data within the context of the international climate negotiations around the United Nations Framework Convention on Climate Change, which again is heavily structured by probabilistic simulation modeling and its data outputs discussed in former chapters.

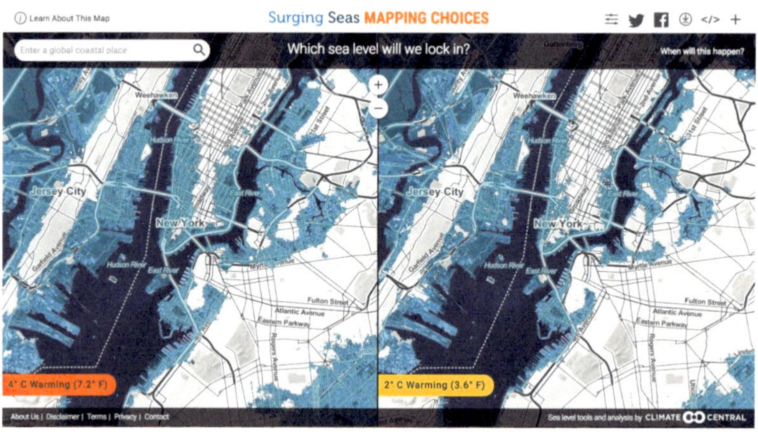

Figure 47: Mapping Choices between a 'due nothing' (4 degrees warming) and a 'transformation' (2 degrees warming) scenario in the area of New York City, USA.[113] Source: Climate Central Website

113 https://choices.climatecentral.org, retrieved on August 9, 2018.

Risk Zone Map

Mapping choices is only one of many manifestations of the Surging Seas imaginary. A second variation is the Risk Zone Map shown in Figure 48. Here, users are not confronted by an explicit visual argument as in Mapping Choices but (literally) build their own image of the future by exploring different scenarios and thematic perspectives.

Figure 48: Climate Centrals Risk Zone Map, here depicting a sea level rise (SLR) scenario of 2 m within the area of New York City, US.[114]
Source: Climate Central Website

The most prominent control element in the dashboard of the map is a symbolic water meter representing different levels of flooding. One can use the water meter as a slider, moving the water meter up and down, hereby flooding the coastal areas with the blue water layer. The spectrum of possible scenarios represented by the scale of the meter

114 https://ss2.climatecentral.org, retrieved on August 9, 2018.

stretches from zero to thirty meters of flooding.[115] The scenarios are subject to considerable interpretative flexibility. One might read the flooded map as a scenario of a distant future with SLR triggered by climate change. It can equally represent a scenario in the near future of inundation caused by extreme weather events, such a hurricane or tsunami.

> Water level means feet or meters above the local high tide line ("Mean Higher High Water") instead of standard elevation. Methods described above explain how each map is generated based on a selected water level. Water can reach different levels in different time frames through combinations of sea level rise, tide and storm surge. Tide gauges shown on the map show related projections (see just below).
> The highest water levels on this map (10, 20 and 30 meters) provide reference points for possible flood risk from tsunamis, in regions prone to them.[116]

Such a 'multi-designation' of scale would not necessarily qualify under the rules of scientific data visualization. Here, however, it allows for an amplification of the argument, while, at the same time, reducing the complexity of navigation.

Driven by open elevation data

We can trace the becoming and further development of the Surging Seas program and its mapping artefacts by making use of the stored historical website snapshots of the Internet Archive and its Wayback

115 The water meter exemplifies the ambiguities of the representation strategies used in Surging Seas: The representation does not adhere to scientific norms in visual representation, in the sense that it packages two different kinds of stories into one Y axis. Zero to 3 m represents scenarios for mean SLR, while 3 to 30 m represents flood risks due to tsunamis triggered by climate change.

116 https://ss2.climatecentral.org/#waterlevel, retrieved on May 3, 2019.

Machine.[117] The analysis shows how the Climate Central's maps are interwoven with the making of high-resolution elevation data produced by numerous actors but coordinated and make available by the NOAA. The highly detailed flood maps of Climate Central would not have been possible without this comprehensive vertical mapping effort within the United States. A description (2015 historical snapshot retrieved from the Wayback Machine) of the initial elevation data available for the maps reads as follows:

> For the elevation data behind our maps, we used the National Elevation Dataset (NED), a product of the U.S. Geological Survey. The NED divides the contiguous United States into a grid of tiny, roughly square cells covering its full area. For each of the millions of trillions of cells, NED provides coordinates and an estimate of average elevation. We used the highest-resolution edition of NED that has full coverage of the coastal contiguous U.S. Cells are approximately thirty feet (ten meters) on a side; this is the finest resolution data publicly available with such extensive coverage.[118]

One does not need extensive expertise in remote sensing technology to consider that such a resolution is barely sufficient for the creation of a robust and detailed three-dimensional picture of the Earth. Even more so, the practices and technologies used to develop the NED dataset

117 Wikipedia entry of the Wayback Machine: "The Wayback Machine is a digital archive of the World Wide Web and other information on the Internet. It was launched in 2001 by the Internet Archive, a nonprofit organization based in San Francisco, California, United States." Last retrieved on July 15, 2019, via https://en.wikipedia.org/wiki/Wayback_Machine. Official website and background information: https://archive.org/about/.

118 Historical snapshot of the Climate Central website captured on January 26, 2015. and retrieved by the Wayback Machine on July 15, 2019 via: https://web.archive.org/web/20150126135901/http://sealevel.climatecentral.org/research/methods/mapping-low-coastal-areas.

varied considerably, as the scientists of Climate Central point out in one of their scientific papers:

> Data sources of varying quality underlie the NED in a complex spatial patchwork, with varying consequences for vertical error from place to place [...]. (Strauss et al. 2012: 4)

Nevertheless, the researchers had done their best to deal with the limited data quality using various interpolation techniques. However, Climate Central also knew that the problem of poor vertical data quality would soon be a matter of the past, due to ongoing vertical mapping efforts by the NOAA. These LiDAR datasets constructed and made available by the NOAAs Digital Coast team enabled Climate Central to improve its elevation models and SLR mappings to amplify their Surging Seas program and produce a number of cartographic artifacts of persuasive detail and quality, such as the Risk Zone Map discussed earlier. Climate Central acknowledges the relevance of this improvement in data quality on a background information page describing mapping methods:

> Improved elevation data: Our 2012 analysis used the best available national coverage elevation dataset at the time. This analysis uses far more accurate laser-based (LiDAR) elevation data.[119]

This text passage triggers a crucial limitation of the Surging Seas assemblage: Its future imaginary, being essentially driven by data, strongly reflects the data divide (Gurstein 2011: 5f) discussed in Chapter V. The million dollar program Digital Coast only produces and makes available data for the coastal areas of the United States, while many of the most vulnerable places threatened by SLR are located

---

119 http://sealevel.climatecentral.org/maps/science-behind-the-tool, last retrieved on May 3, 2019.

within developing countries. This problem of insufficient data quality outside the US is also mentioned on Climate Central's website:

> Outside of the U.S., very little lidar data is available. Instead, we use radar satellite-based data collected from NASA's Shuttle Radar Topography Mission (SRTM). This elevation data covers nearly the entire populated world, but is less accurate than lidar. SRTM's pixel resolution is lower, and in areas of dense urban development and vegetation, SRTM tends to overestimate elevation. Recent work also suggests that SRTM usually underpredicts exposure from sea level rise and coastal flooding. Outside the U.S., our flood maps should therefore be seen as likely lower bounds on the extent of potential inundation for each water level.[120]

Figure 49: Scenario of 1 m SLR in St. Louis, Senegal (left), and Lower Manhattan, US (right). Source: Climate Central

120 http://sealevel.climatecentral.org/maps/risk-zone, last retrieved on May 3, 2019.

The comparison of the risk zone mapping for the vulnerable city of St. Louis in Senegal and Lower Manhattan in the US (Fig. 49), illustrates the data divide within the imaginary of Surging Seas convincingly. As a matter of fact, the data divide is even stronger than the maps suggest. The resolution of the elevation model available was higher than that shown in Surging Seas but had to be lowered *ex post* for data privacy reasons (similar to the agricultural maps discussed in Chapter III). The data divide is even more amplified through the mobilization of the Surging Seas mappings within digital media platforms, which is discussed in the following.

Googlelization of the future

A variation of the mapping imaginary is a series of YouTube videos. Climate Central developed a so-called 'extreme scenario 2100' within its Surging Seas program implemented through a content layer in Google's geocoding language Keyhole Markup Language (KML)[121]. This content layer can be rendered as a compelling three-dimensional landscape in Google Earth, showing a potential worst-case scenario of a future with SLR triggered by climate change. The animation does not only give users the possibility of discovering the flooded world on Google Earth (Web), the altered framework has also been operationalized as a video maker for Climate Central itself, namely, to develop tailor-made video animations of the extreme scenario for major cities. The videos are created with the built-in fly-through function of Google Earth, simulating a perspective from a low flying airplane shooting video footage.

121 Description on Google website: "KML is a file format used to display geographic data in an Earth browser such as Google Earth. You can create KML files to pinpoint locations, add image overlays, and expose rich data in new ways. KML is an international standard maintained by the Open Geospatial Consortium, Inc. (OGC)." Retrieved on June 4, 2019, via https://developers.google.com/kml/.

Climate Central produced many (about 30) of these flythrough videos to show potential SLR in New York, London, Osaka, Buenos Aires and other major cities worldwide. Above, you see the animation of London with a simulated SLR according to a 4 °C global warming scenario. Similar to *Mapping Choices,* the caption "4 °C" puts the imagery into perspective and attributes the visible flood to the invisible phenomenon of climate change. The 'worst-case' (4 °C) and 'best-case' (2 °C) scenarios are shown here one after the other to give an impression of the difference between the two possible futures.

Figure 50: YouTube video by Climate Central: Mapping Choices Global Tour[122]
Source: Climate Central/YouTube

Browsing through the whole set of Climate Central's flythrough videos on YouTube, it is remarkable that these only show particularly popular places, major cities. It almost feels like a tourism advertisement, were it not for the areas tainted in dark blue. The videos fly through the main attractions of the city, showing many of them flooded. The message is

122 https://www.youtube.com/watch?v=VeXwN0ju888, retrieved on May 3, 2019.

clear: We will potentially lose a lot of the places we love to travel to. These places could vanish forever. While the Risk Zone Map struggles with a data divide between the US and the world, the Google Earth extreme scenario amplifies but also alters the composition of the divide. The Google Earth scenarios have been enabled through the decision of the NOAA to provide their datasets formatted as KMZ files. This choice of the KMZ format can be understood as an element of a wider strategy of the NOAA (and other US-American agencies) to mobilize the power of Silicon Valley to address matters of social concern. In a presentation I witnessed at Telegrafenberg in 2016, the NOAA's chief scientist Rick Spinrad called these companies the 'AMIGOS' of the NOAA, loosely referring to companies such as Amazon, IBM, Google and Oracle.[123] Thanks to the provision of LiDAR data as KMZ files, it was possible to depict the SLR scenarios of Surging Seas within three-dimensional landscapes powered by Google Earth. In this version of the mapping assemblage, Surging Seas leaves its focus on the United States and represents itself as an international initiative, visualizing SLR within multiple major cities around the world. However, the choice to 'go Google' also introduces a new data divide, namely between major cities and rural areas. As for the Risk Zone Map, the possibilities and limits of visualization are driven by the data available. Accordingly, the Google Earth scenario deals quite well with the rendering of a few major cities in emerging economies, thereby, lowering the bias towards Western cultures in the future imaginary. This includes visualizations of Buenos Aires, Rio de Janeiro, Osaka, Hong Kong, Dubai and Tokyo. However, similar to case of the Risk Zone Map, it is impossible to depict SLR at a reasonable resolution in especially vulnerable cities of the Global South, such as Dhaka (Bangladesh) and St. Louis (Senegal).

Aspects of GeoGooglization have been highlighted by Tristan Thielmann and others in the article *Dwelling in the Web: Towards a*

---

123 Presentation by Rick Spinrad at the GFZ on September 16, 2016.

Googlization of Space (Thielmann et al. 2012). On the one hand, Google technologies and services influence the way we ('users') depict, perceive and navigate spaces, what Thielmann and colleagues characterize as 'frontend GeoGooglization' (ibid.: 27ff). On the other hand, Google's localization technologies have enabled wide-ranging profiling and commodification of users, countries, cultures and communities, thereby inducing a 'backend GeoGooglization' (ibid.: 34ff). The example of Surging Seas suggests that (Geo-)Googlization has actually intensified since the publishing of the article (2012). Not only are geomedia user practices, visual aesthetics, and software dominated by Google, but so are the most common data standards (KML). Moreover, the Google Earth imaginaries are then embedded as flyover video animations on YouTube, which is again part of the Alphabet family. We have seen in other chapters of this study that Googlelization might also include other aspects, such as dominant programming languages (Python) and valuation in science (Google Scholar).

Mobilizing allies

Moreover, the Risk Zone Map mashes up the inundation data with other socioeconomic data and indicators, such as social vulnerability, income, population density, ethnicity, property values and cultural landmarks. By choosing 'property values,' for example, one can identify real estate at risk of flooding and SLR. Red areas in New York City in the snapshot shown in figure 50 designate the highest values of over 100 million USD per acre. On the one hand, this constitutes information to be considered for those owning or renting properties in these areas. This is especially relevant in a country which is run by a person owning vast numbers of such properties (i.e. President Donald Trump). On the other hand, for those personally affected, it has entertaining value to browse through the expensive beach houses of the rich which might be drowning in a foreseeable future. The maps generally show that SLR and flooding often harm those communities who are already vulnerable for other reasons (ethnic minorities, the poor). Equally, it shows that wealth does not necessarily protect from climate change. Bruno Latour showed in his characterization of maps as immutable mobiles that these artifacts are so powerful because they enable the effective mobilization and assembly of allies in one place (1988: 23). Surging Seas is a good example of this power of mobilization, in this case, drawing together various agents behind the causes of Climate Central. These agents involve collective human actors, such as representatives of ethnical minorities, the poor, the socially vulnerable, disaster prevention, real estate agencies, federal and municipal authorities, and politicians. They also include nonhuman elements, such as datasets (elevation data, socioeconomic indicators) and superior project funding.

Figure 50: Risk zone mapping of vulnerable real estate.
Source: Climate Central Website

(Re)Purposing maps

The technical mutability of the maps (e.g. Mapping Choices vs. Risk Zone Map) and the interpretative flexibility of meaning (SLR vs. flood) enable Climate Central to position its imaginary effectively in various kinds of discourses and debates. While Climate Central has initially been a well-established actor within the liberal community of the climate debate, Surging Seas enabled the NGO organization to escape from there and to reposition itself as an actor beyond the climate bubble.

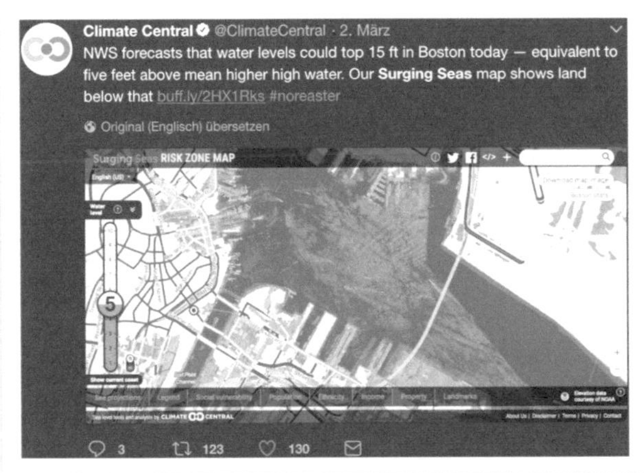

Figure 51: Tweet by Climate Central on March 2, 2018: "NWS forecasts that water levels could top 15 ft in Boston today – equivalent to five feet above mean higher high water. Our Surging Seas map shows land below that buff.ly/2HX1Rks #noreaster"[124]

The best example of this strategy is the repurposing of the Risk Zone Map to comment on ongoing events of public concern. Severe bomb cyclones hit the Eastern coast of the United States in March 2018 and caused severe flooding and damage in Boston and other major cities.[125] Before the event, Climate Central used its maps to produce a short-term (severe) weather forecast per Twitter, as shown in Figure 51. The risk zone mapping stack available was used to simulate flooding expected

124 March 2, 2018.

125 https://www.businessinsider.de/noreaster-bomb-cyclone-boston-floods-2018-1?r=US&IR=T, retrieved on May 3, 2019.

to take place during the next very day of the posting. In so doing, the spatialities available were temporally situated and attributed to events experienced in the present and reported in national and international media.

Co-creating digital publics

As we can see from the preceding examples, the mappings of Surging Seas are not only presented on their own website but effectively embedded and positioned within digital (social) media platforms. In so doing, Climate Central manages to create digital publics that are vividly discussing aspects of the cartographic imaginaries presented. An example are the post-video discussions triggered by the movie *Mapping Choices: Global Tour*[126] on YouTube. The clip shows a fly-through animation depicting popular sites of major cities flooded with water, caused by climate related SLR. The video is called Global Tour, as it is basically a 'best of' of other Climate Central videos, each focusing on one major city. The video is featured in *The Daily Conversation,*[127] a popular channel featuring mini-documentaries about a variety of topics. Glimpses of the comments show that most interactions do not expressively relate to the content of the map (flooding scenario X at location Y). Instead, the cartographic artifact is used as a container and discursive device for discussions of ongoing events and statements of political and ethical views (similar to the *anchoring device* discussed in Chapter III). At the same time, the comments can serve as digitally accountable proof for public interest in the Surging Seas maps and the work of Climate Central more generally. For Noortje Marres, participation in digital societies becomes a resource and valuable good:

126 https://www.youtube.com/watch?v=ekhLHzxc92U&t=8s, retrieved on April 4, 2019.

127 https://www.youtube.com/user/TheDailyConversation, retrieved on April 4, 2019.

> This valuation of participation takes specific forms in digital arrangements, and includes the production of data – with X users signed up, we gain access to X number of profiles – and metrics about participation – so many unique visitors, so many comments – and the development of communicative strategies for market-, audience- and brand-making – the website as community; the identification of targetable influencers. (2017: 151)

The valuation of participation is *per se* not a new phenomenon – more traditional participative arrangements, such as focus groups and stakeholder consultations, have always generated a significant amount of data and inhibited opportunities for publicity. However, as Marres argues, the deployability of participation is particularly notable in digital societies, where "participation data and metrics are not only taken up by experts, but are rather made available to a variety of third-parties and users themselves, thus identifying their deployability" (ibid.: 152)

Such deployment of participation has been explicitly operationalized by Climate Central as an excerpt from the Surging Seas website shows:

> Our Outreach is What Inspires Us.
> Press coverage of the sea level program's work so far totals more than 5,000 stories. [...] More than 60 federal agencies and offices, and more than 100 state agencies, county offices, and city offices each have accessed our tools, plus numerous businesses, nonprofit organizations, and educational institutions. [...] Our sea level content has received more than 10 million page views.[128]

Accordingly, the extraordinary power of the Surging Seas imaginary is not only caused by its ability to provide visual proof of future flood risks but also effectively valuate the digital publics mobilized through the imaginary. These representations of publics' participation typically

128 http://www.climatecentral.org/news/coverage-of-surging-seas-inundates-the-nation, retrieved on April 2, 2019.

come in the form of numbers. According to Marres, such metricization constitutes a further feature of participation within digital societies:

> Digital media technologies specifically enable the monitoring, measurement and analysis of participation, and these practices are critical to their transformative effects on society. (2017: 156)

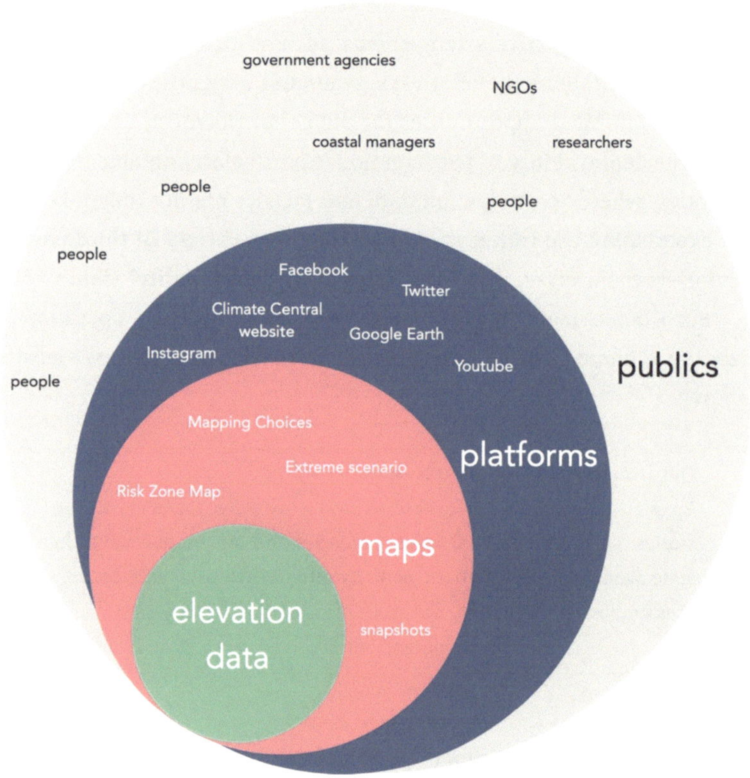

Figure 52: The socio-technical imaginary of Surging Seas, as constructed by Climate Central and captured by my own (digital) research devices. Source: My own visualization

Digital engagements of publics always leave traces, which are captured and transformed into numeric form (e.g. number of likes, mentions, views, comments). In so doing, these traces of participation can be attributed to an actor (e.g. Climate Central), to an assemblage of artifacts (e.g. Surging Seas) or to the relevance of a matter of concern (e.g. SLR). Marres shows that such a valuation of participation within digital societies can be further associated with two characteristics: It is increasingly enabled and mediated through digital technologies and it operates along politics of metricization: "[...] as digital media technologies proliferate across social, political and public life, they have become ever more visible and notable elements in the doing and staging of participation." (ibid.: 153)

Fluidity and viscosity of digital technology

During the analysis of Surging Seas, I was again confronted with the issue of fluidity that has been addressed in earlier chapters. How to account for fluidity in shape and meaning? How to identify and organize fluid elements to make them graspable for analysis? What is changing and what stays the same? And what are the decisive elements, moments, constellations, deciding whether something changes or not? Marianne de Laet and Annemarie Mol (2000) have addressed some of these questions in their case study of the Zimbabwe Bush Pump, a common water pump used in rural neighborhoods. Challenging the idea of immutability in technology, they introduce the concept of *fluid technology* in order to account for aspects of instability, flexibility, temporality and fluidity:

> The Zimbabwe Bush Pump has existed for more than half a century, but it has not remained the same. It is not an immutable but a changeable object, that has altered over time and is under constant review. (ibid.: 228)

As a matter of fact, the *fluid technology* concept has often been evoked along the description of digital phenomena, including online communities (Faraj et al. 2011), the use of mobile phones (Herold et al. 2013), laptops in developing countries (McArthur 2009) and education 2.0

settings (Selwyn 2012). The *Fluidity* of technology is also an aspect discussed on various occasions in *Time for Mapping*, an anthology addressing temporal perspectives on digital mapping endeavors (Lammes et al. 2017). Moreover, fluidity is explicitly mentioned in the context of sustainable energy and climate modeling by anthropologist Kirsten Hastrup. Referring to de Laet and Mol's article, she argues that

> [...] technologies of weather and wind, whether designed to harness, to measure, or to mitigate their potential, are fluid objects in the sense described here. They will not work if too rigid, because the weather-worlds in which we live are not rock-solid, but the opposite. The fluidity of the objects is remarkable also when we consider the computer technologies, now capturing the fluidity and complexity of the climate system. (Hastrup 2013: 18f)

The issues discussed by these authors show some similarities to our case of Surging Seas and, more generally, to the phenomena discussed within the book. Against this background, the following paragraphs will elaborate on the arguments of de Laet and Annemarie Mol and consider some aspects more in detail. The hypothesis is that the Bush Pump can help us to understand and describe particularities of digital technological phenomena, using the assemblage of Surging Seas as an example.

De Laet and Mol begin by describing the solid mechanical elements of the pump, including a water discharge unit, a steel pump stand and a lever: "Of course, all this is held together by nuts and bolts" (2000: 228f). For illustrative purposes, they also provide a schematic image of the pump taken from an available instruction manual. However, as de Laet and Mol highlight, this particular shape does not make a pump yet. The pump is also defined by its hydraulic principles that make it *work:* "The hydraulic forces draw water from deep wells to the surface" (ibid.: 230) And the particular way in which the hydraulic principles work make the Zimbabwe Bush Pump part of certain family of pumps, those with a "lever activated lift pump mechanism" (ibid.). Within this family, it is the Bush Pump's capacity that makes it specific and different from all other pumps of this family: "The Bush Pump's strokes are more efficient and powerful than those of most other lift pumps [...]." So, the Bush Pump is specific, but the characteristics that distinguish it from each of these also tend to be shared with one or more of the others. "For the Bush Pump, *'being itself' means that it is continuous with a number of others*" (ibid.: 230f, emphasis in original).

Does the characterization of the pump end here? De Laet and Mol would object: "[...] there is a problem, for when it's unloaded from the truck the Bush Pump yields no water. None whatsoever. It is not a pump" (ibid.: 231). For the pump to yield water, it has to be assembled and installed properly into concrete headworks and equipped with a casing. In this way, "it becomes a source of pure, fresh, *clean* water. And so the Bush Pump turns out to be a technology that provides not just water but also health" (ibid.).

Moreover, no technology operates in the void. The bush pump has to collaborate with others to be functional and successful: A tube well drilling device (ibid.: 233) and a community of villagers:

> The pump is nothing without the community that it will serve. In order to be a pump that (pre)serves a community, it not only needs to look attractive, have properly fixed levers and well-made concrete aprons, it must also be capable of gathering people together and of inducing them to follow well-drafted instructions. (ibid.: 234 f)

More than that, the pump does not only serve communities. It helps to hold them together. "As it helps to distribute clean water, it also builds the nation" (ibid.: 235).

To summarize this, we can identify two characteristics of technology that determine its *fluidity:* Firstly, it can be many things at the same time and its boundaries are not solid and sharp. It is equally a mechanical object, a hydraulic system, a provider of water and health, a community servant and a nation-building apparatus. Secondly, its components may change over time but its essence remains stable. The ability to handle temporary break-down and deploy alternative components is a source of strength in fluid technology (ibid.: 253).

The fluidity of Surging Seas

As we can see, the concept is itself fluid and subject to interpretative flexibility. This fluidity may also have been a source for its wide applicability in STS and beyond, with heterogeneous studies all referring to the peculiar technology of the Bush Pump. In the following, we can consider the different aspects of fluidity in dialogue with the empirical case of Surging Seas.

Many things at the same time

Similar to the pump, Surging Seas can be characterized as multidimensional, with these dimensions being interrelated and dependent on each other. If the pump is a mechanical object, a hydraulic system, a provider of water and health, a community servant and a nation-building apparatus, which are the different layers of entities working in Surging Seas? First of all, it is a mapping stack, an entanglement of software elements and aesthetic conventions enabling the depiction of the data on visually appealing digital maps. Secondly, Surging Seas is an informational node, drawing together data and information about probable future SLR and flood risks. Thirdly, it is an sociotechnical imaginary (Jasanoff/Kim 2015), mediating a specific perspective on the future – a future framed essentially as risky territory. Fourthly and building on that, Surging Seas also proposes certain ways of handling this risky future, which are again framed by the propositions that maps tend to make (Wood/Fels 2008). Fifthly, Surging Seas is a conscription device (Henderson 1991), capturing and interrelating a variety of publics for the cause of Climate Central (i.e. reducing GHG emissions and building more climate-resilient communities and environments). Sixthly, it is an anchoring device (see Chapter III) for debates regarding the propositions of the map. Finally, similar to the pump in Zimbabwe, Surging

Seas is increasingly engaging in nation-building. It aims at uniting different parties with potentially conflicting views behind a common challenge and vision to deal with risky futures.

Many things at different times

From time to time, the Zimbabwe bush pump will fail to deliver some of the services it usually promises – water, clean water or health for a specific community. Bolts may fall out of the bucket, the community may be too small (maintenance) or too big (capacity) for the pump, or the water quality is insufficient to meet standards for drinking water. However, this does not make the pump essentially fail. It can be repaired quite easily, spare elements are manufactured and available at multiple locations, and communities may well find temporary ways of handling poor water quality. The pump as a technology is fluid because it has incorporated the possibility of its own breakdown and the flexibility to deploy alternative components. It continues to work to some extent "even if some bolt falls out or the user community changes" (De Laet/Mol 2000: 253).

Nevertheless, we may ask: What are the elements of the bush pump as a technology that are the most difficult to exchange? Which are the elements whose dysfunction or absence is the most prone to breakdown or perceived as a 'failure' of the technology as a whole? At one point, de Laet and Mol address this question in their article:

> Spokespeople in Zimbabwe pointed out to us that the continuation of its manufacture has been a fragile element in the working of the Zimbabwe Bush Pump 'B' type. For a long time it seemed as if it might be its most fragile element – and if this was the case, then it was precisely because it is the least fluid. (ibid.: 247)

In this reading, a technology consists of 'fluid' and 'less fluid' elements, where 'least fluid' elements are considered as 'fragile' (e.g. the manufacturing system). In the following, I would like to propose a reinterpretation and shift of focus in the consideration of fluidity in technology.

Is digital technology essentially fluid?

I would argue that the concept of fluidity in technology is often used not to describe the relationship of fluid and less fluid elements but to circumscribe characteristics such as 'complexity,' 'invisibility,' 'intangibility' or 'lacking ascertainability' of contemporary technology as a challenge for human cognition. Thomas Sutherland investigates the history and meaning of 'flow' and 'fluidity' metaphors for cartography in his article in the anthology *Time for Mapping* and comes to a similar verdict:

> So why, we must ask, has this trope become so popular? Why is the term so frequently, uncritically and off-handedly deployed in the social sciences, and especially within the practices of mapping that have grown in dominance within these disciplines? The simple answer is probably to a large degree the correct one: the image of fluidity is an effective metaphor for the way in which network-driven distribution channels are able to transmit goods, information and even people at *rates and speeds* that make them effectively unthinkable by the human intellect alone, particularly when attempting to represent these movements in a visual manner. (2018: 191, emphasis in original)

In my opinion, it was also this synonymization of fluidity with 'ungraspability for the human eye' that has triggered the popularity of the fluid technology concept for the description of digital and spatiotemporally distributed phenomena, for example, not only algorithms, databases and computer models but also large infrastructures, such as Astrid Hastrup's wind farms. The term *fluid technology* may, therefore, seduce researchers to be content with the determination of fluidity and the identification of fluidity elements[129] which can be illustrated with an example from Surging Seas. In the latter, many elements show characteristics of fluidity in the sense described by De Laet and Mol, one of them being datasets describing socioeconomic indicators:

> Our 2012 research assessed land, population and housing vulnerable to sea level rise and coastal flooding. This research assesses over 100 additional variables, including socially vulnerable population, populations by racial and ethnic group, property value, roads, rail, airports, power plants, sewage plants, hazardous waste sites, schools, churches, and hospitals.[130]

As a matter of fact, the integration of socioeconomic indicators differentiates Surging Seas from other mappings of SLR which became popular around the years 2007–2010. All these cartographies entailed mainly three categories of information: Climate scenarios, inundation and infrastructure. The master narrative (Star 1999) of these flood maps is simple accordingly: Climate change will cause severe inundation in coastal areas and damage to buildings, roads and other infrastructures. Early SLR maps sometimes explicitly indicated 'infrastructure at risk,' such as healthcare facilities, schools, police and fire stations, waste-water treatment plants and nature conservation areas (see Fig. 53). These mappings (consciously or unconsciously) mediated a blunt narrative of risk. And as Ulrich Beck once declared: "Risk is a modern concept. It presumes decision-making. As soon as we speak in

129 The same critique can be made for the prolific use of the terms 'heterogeneous' and 'mobile' in STS, which is equally problematic in my opinion.

130 http://sealevel.climatecentral.org/maps/science-behind-the-tool, retrieved on December 11, 2020.

terms of 'risk', we are talking about calculating the incalculable, colonizing the future"[131] (Beck 2002: 40).

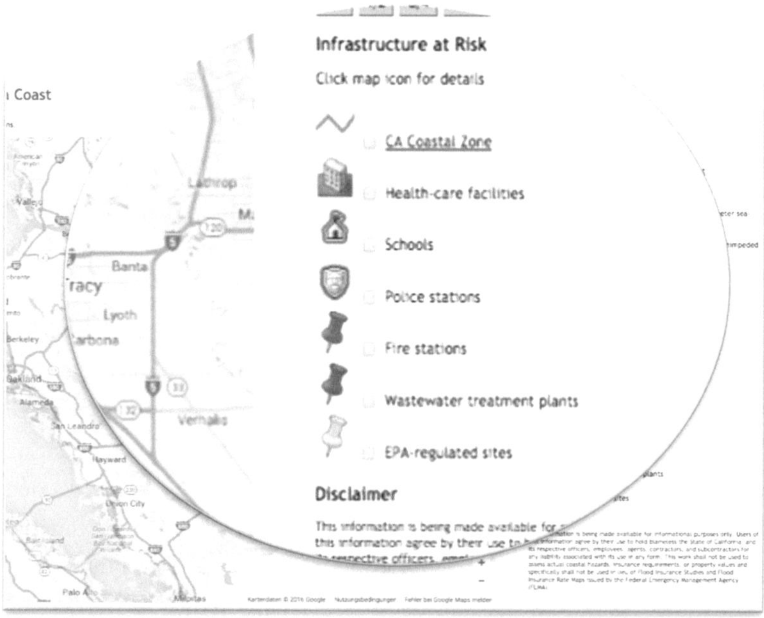

Figure 53: Early SLR mapping by the Pacific institute, detail enlargement of 'infrastructure at risk.' Snapshot taken March 2016 (Heberger et al. 2009 for publication, map no longer available)

On the one hand, 'infrastructure at risk' calls for levees and other territorial protection measures. On the other hand, it makes people check on the map whether they are affected by climate change impacts or not. If not, they may discard climate change as a relevant issue in their life. Surging Seas' association of inundation data, climate change scenarios and socioeconomic indicators shift the master narrative of SLR

131 It should be noted that Beck made this comment in the context of his assessment of counterterrorism measures after 09/11. Despite the very different context, this evaluation oscillates with the narrative of these mappings.

mappings. It mediates SLR and flood risks as issues of national concern, and pretext for solidarity. Similar to the bush pump in Zimbabwe, the SLR mappings engage in community- and nation-building. They put the issues of climate change and flooding into perspective with local and national matters of concern, such as racial discrimination and poverty. In so doing, they call for solidarity between regions, communities and citizens. While socioeconomic indicators alter the narratives of the mappings and amplify their significance and agency, they are not conditional for the whole program. They are fluid elements in the assemblage which can be added or removed rather easily without causing a breakdown or a perception of failure. If they are present, they flow through the assemblage without strongly changing its elementary structure. They are no essential dependencies of the assemblage. By contrast, there are elements that are central for the functionality, performance and success of Surging Seas. I would argue that we can characterize these phenomena as *viscous elements*. They are fluid but less fluid than other elements in an assemblage. They are *viscous* in the sense of *sticky* and *persistent* in an ever-changing network space.

Viscosity in technology

Viscosity in fluid mechanics is a measure of the internal resistance of a liquid to flow:

> The term viscosity is commonly used in the description of fluid flow to characterize the degree of internal friction in the fluid. This internal friction, or viscous force, is associated with the resistance that two adjacent layers of fluid have to moving relative to each other. Viscosity causes part of the fluid's kinetic energy to be transformed to internal energy. (Serway 1996: 427)

I would argue that we can characterize the behavior of digital technologies as a flow of different liquids. Everything may flow together but the different liquids are flowing with varying intensity – they are more or less fluid (and viscous vice versa). Accordingly, the behavior of some elements in digital technologies may be characterized as highly fluid (data flowing through a machine learning process), and others clearly are not (e.g. hardware, server rooms, cables, screens). Elements that are viscous are in between. They are persistent and tend to affect the flowing behavior (velocity, direction) of other elements.

I would argue that the reading of mapping as practice overplays the aspect of mutability in digital maps by black boxing their underlying material base, which is rather stable. We may gain the impression of a constant unfolding of a mapping by observing people's use of a navigator app: The map changes as the app is used and so does its interpretation by the user. However, the material elements in this situation remain rather immutable the whole way through: The smartphone, the app, the operation system, the geodataset, the design classes and attributes, and the servers of the app provider. The approach of this study gives more weight to the material elements and assemblages underlying the map. It abandons not only a focus on the visual surface of lines, points and areas, but also on practice (making and using mappings), and digs deeper into codes, data flows and entangled infrastructures.

Web software development provides multiple devices that can be repurposed for inventive methodologies. We can use the developer tools available in web browsers, for example, to uncover the source code of a map (or website) and identify key dependencies, such as libraries and datasets. Toggling over the Risk Zone Map, the inspector tool reveals the *Leaflet* software library as a main structural component of the mapping stack (see Fig. 54). Leaflet is an open-source JavaScript library used to build web-mapping applications. Along with OpenLayers[132] and the Google Maps API, it is one of the most popular JavaScript mapping libraries and runs in the background of websites and platforms

132 https://openlayers.org/, retrieved on May 3, 2019.

such as FourSquare, Pinterest, Flickr, Meetup, Craigslist, the Wikipedia mobiles applications and various online media providers.

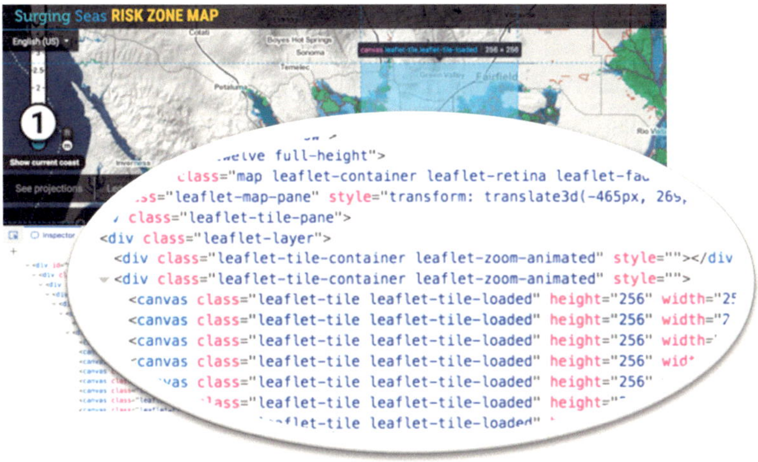

Figure 54: Source code inspector in the web browser Firefox.
Source: Own screenshots

While Leaflet enables the integration of various data sources, it is mostly used in combination with Open Street Map data. It is a free alternative to proprietary services, such as the Google Maps API[133] and ArcGIS online. First released in 2011, it supports most mobile and desktop platforms.[134]

---

133 Wikipedia entry: "In computer programming, an application programming interface (API) is a set of subroutine definitions, communication protocols, and tools for building software." Retrieved on May 3, 2019, via https://en.wikipedia.org/wiki/Application_programming_interface.

134 Wikipedia entry: "Cascading Style Sheets (CSS) is a style sheet language used for describing the presentation of a document written in a markup language like HTML. CSS is a cornerstone technology of the World Wide Web, alongside HTML and JavaScript." Retrieved on May 3, 2019, via https://en.wikipedia.org/wiki/Cascading_Style_Sheets.

Leaflet allows web developers without comprehensive skills in geoinformatics to display interactive web maps hosted on a public server, with optional tiled overlays. It can load feature data from GeoJSON files (e.g. from the Open Street Map API), style it and create interactive layers, such as markers with popups when clicked.[135] It is extremely lightweight and has no external dependencies. Leaflet was developed by Ukrainian software engineer Vladimir Agafonkin, together with a strong developer community that is spread worldwide. To date (July 5, 2019), 618 developers had contributed 6,754 times ('commits') to the code of Leaflet on Github,[136] updating, debugging, improving, complementing and documenting it with continuous persistency.

How may we conceptualize the role of Leaflet for Surging Seas? It is an essential element in the assemblage because it enables the mobilization of the elevation models and their depiction as flood maps within the ecology of the web. In fact, there would be other solutions to achieve this outcome but they have several downsides. Proprietary services, such as Google Earth Engine and ArcGIS Online, provide similar functionalities as Leaflet, but their use would be associated with considerable financial costs. This financial aspect is the reason why Leaflet is so popular among NGOs, online journalists and start-up companies. Moreover, using proprietary services would make Climate Central dependent on the functionalities of their services. By contrast, Leaflet, as an open Javascript library, is endlessly modifiable and combinable with other elements. This characterization evokes a comparison with the trope of the *immutable mobile* by Bruno Latour. Referring to maps in particular, Latour lists a number of characteristics that constitute the power of immutable mobiles: They are mobile, they are immutable when they move, they are made flat, their scale may be modified

135 https://wiki.openstreetmap.org/wiki/Frameworks, retrieved on May 3, 2019.

136 https://github.com/Leaflet/Leaflet, retrieved on May 3, 2019.

at will, without any change to their internal proportions, they can be reproduced and spread at little cost, so that all the instants of time and all the places in space can be gathered in another time and place, they can be reshuffled and recombined, it is possible to superimpose several images of totally different origins and scales, they can easily become part of a written text and their two-dimensional character allows them to merge with geometry (Latour 1988: 19f). Several authors (Abend 2018; Lammes 2017; Perkins 2014) have investigated aspects of (im-)mutability in contemporary digital geomedia. Sybille Lammes, for example, has argued that cartographic images in digital mapping enterprises might have become mutable, but the "digital map as a network of control" remains stable:

> Although the image itself may have become mutable since the advent of digital mapping, the digital map as a network of control is still immutable for the map source is stored in a database (e.g. Google Maps) that is not easily transformable and operates according to set rules. (Lammes 2017: 1030)

I generally agree with this interpretation. The GIS powering CIO, discussed in Chapter III, is a good example of such stability of the map-underlying "network of control." Nevertheless, the case of Surging Seas and the operationalization of Leaflet suggests a slight alteration of this characterization. Leaflet seems to be the element which *affords* many of the characteristics described by Latour: It enables maps to be mobilized within the web, three-dimensional elevations to be flattened on a two-dimensional space, to be scaled and modified without any change in their internal proportions, to be reproduced and spread at virtually no cost, to be reshuffled and recombined with multiple information layers, and to be supplemented by written text and positioned within (social media) fora of debate. Is Leaflet the immutable mobile then, enabling the mobility of the Surging Seas maps? I would argue that this is not the case. Libraries such as Leaflet are updated frequently by their large community of contributors in order to run smoothly with the pace of the web. Similar to the Zimbabwe bush pump, their relative fluidity is the key to their internal stability. Nevertheless, one might miss the point to characterize Leaflet as a 'fluid technology.' Instead, it can be portrayed as a sticky, persistent, *viscous object* within a fluid assemblage of technological elements.

Conclusion

I have become a 'Scientist for Future' myself in explicit and implicit ways during the last few years. I signed and supported the birth-giving statement of the 'Scientists for Future' movement declaring that *The concerns of the young protesters are justified. A statement by Scientists for Future concerning the protests for more climate protection* (Hagedorn et al. 2019). I have also acted as a mediator, putting proponents of the Scientists for Future movement together with scientists from the Potsdam Institute. During an Open Science fellowship at the Wikimedia Foundation, I collaborated with the library of the Telegrafenberg and with computer scientists to translate the entire publication database of the Potsdam Institute to the open knowledge base Wikidata. In one of the preliminary steps of this translation, we visualized the co-author networks of the PIK scientists, as shown in Figure 55. After a while, I discovered myself as a (very small) node within the network, co-author of the PIK publication *Climate Impacts for German Schools – An Educational Web Portal Solution* (Blumenthal et al. 2016). It was the realization that I had equally become an element of my field and part of my own research data.

It would be wrong to characterize this situation as a mimesis but rather as a shifting of positions of the field and the researcher. 'The field,' as I have described it within this book, had not existed four years ago. While such processes may have started before in other locations, the last four years have been a crucial period in the mutual construction of infrastructures, the mobilization of artifacts and the establishment of practices for (what is increasingly labeled as) 'open science.' I have tried to describe this shift within this book and to reflect on some of the consequences for scientific practice. Equally, I have been a (very marginal) contributor in the construction of this field, experimenting with new ways to 'open' science and to feed in the perspective of STS and ethnography to such processes. The following paragraphs provide a more condensed view of the shift from traditional practices in climate impact modeling to the quest for digital openness.

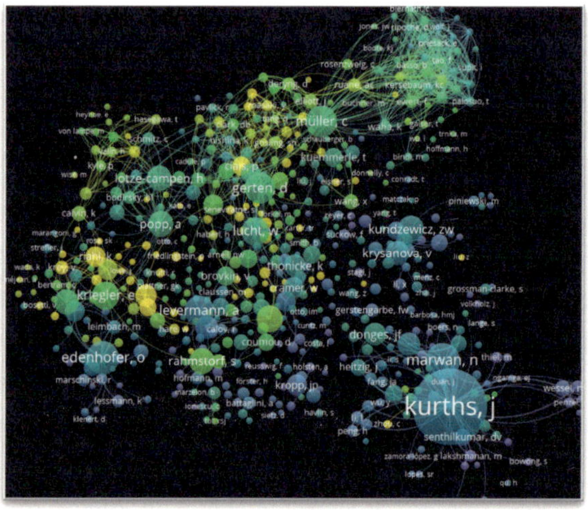

Figure 55: Co-author network of the PIK publication database, visualized with the software VOSViewer.[137] Source: My own visualization

Shifting temporalities in scientific modeling

The reconfigurations documented within this study represent a temporal restructuring of the moments of opening and closure within scientific modeling work. These altered temporal orderings are both

137 VOSViewer: https://www.vosviewer.com/, retrieved on June 14, 2019.

triggered by the mainstreaming of digitally networked infrastructures and socio-political expectations for open science.

Impermeable modeling

Locating moments of openness and closure in traditional climate modeling processes is relatively simple. We can illustrate this by adding to Latour's visualization of the cascade of inscriptions discussed earlier.

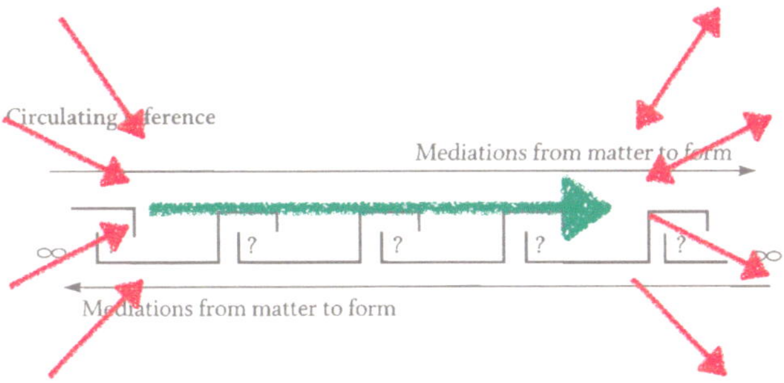

Figure 56: Moments of opening and closure in impermeable modeling.
Source: Latour 1999a supplemented by additional elements

The green arrow represents the transformation of information within the process and time span of a modeling enterprise. Research endeavors in applied research, such as climate impact modeling, are typically triggered by tenders of founding institutions. Scientists try to align their skills and research interests to the demands of the research institutions and frame their proposals as solutions to the problems referred to in the call for projects. Scientists (mostly in teams or consortia) apply for the grants and some of them are funded. Sometimes, new people are hired; more often, existing ones are rededicated to the new project. Scientists gather existing theory, models and data to build upon. Put otherwise, the scientists breath in and seek inspiration from outside of their project team. In Figure 56, this incorporation of information at the beginning of the research process is represented by the one-directional arrows in magenta on the left. After an initial phase of conceptual and organizational reshuffling, the dynamics of scientific processes become more stabilized. The incorporation of outside knowledge is heavily narrowed down to a level perceived as manageable for the researchers, which is represented by the green arrow in Figure 60. This strategy allows for the scientific analysis and knowledge production within the conceptual and methodological boundary conditions of computer modeling. Considering the way in which simulation modeling processes are temporally structured, it is difficult for scientists to incorporate new elements before the simulation rounds are through and the results formally evaluated. Small conceptual changes may have great and unforeseeable consequences for the entire behavior and performance of the model due to the complex mathematical logic and computational architecture of simulation models. Making statements about ongoing modeling processes prior to the formal evaluation is often difficult. Modelers, therefore, often refrain from communicating ongoing research and save representational activities for the final period of the research process. This also prevents them from becoming victims of political attacks from climate skeptic forces who defame climate research as 'fake science.' In Figure 56, this opening up to outsiders at the end of the research process is represented by the magenta arrows on the right. The arrows go both ways, as they include mediation from the inside to the outside but also from the outside to the inside. These interactions between scientists and publics take various forms and include developing scientific papers, producing visualizations, giving interviews to journalists and producing popular science books. On the other hand, the feedback received at the end of the research process may lead to alterations of the models applied, the integration of additional aspects or an alteration of priorities. I call this *impermeable modeling.* The scientific process is sealed within the core period of research. It is only at the end of the epistemic and organizational process that the seal is opened and the newly created knowledge is set free for communication and social negotiation.

Permeability and digital openness

As I have shown in the chapters of this book, the temporalities of interaction within scientific modeling are currently changing. Moments of opening, interaction and feedback with outsiders do not only occur at the end of a modeling process but from the very beginning and throughout projects. As discussed in chapters III – VII, researchers contribute to Github repositories and software libraries, they upload and document vast digital datasets and produce images for their engagement with various stakeholders. Along with these activities, they receive feedback, which explicitly and implicitly informs their work in progress. In other words, modeling becomes more and more *permeable*. The interaction with others within permeable modeling is sometimes happening face-to-face – scientists increasingly speak to social actors, such as journalists, policy makers, teachers and pupils, or more fuzzy categories of 'the general public' or 'non-experts.' In some cases, scientists are comfortable with this new evocation of representation, in other cases, they instead feel overwhelmed by these new expectations. These transformations in the practices of science communication have been widely reflected within the academic literature as conceptual shifts from characterizations of science as an ivory tower, through the public understanding of science, toward public engagement with science and technology (Schäfer et al. 2015).

In this book, I have focused on another transformational shift in scientific practice and communication driven by the rise of digitally networked infrastructures. Nowadays, the mediation of scientific knowledge is increasingly organized through the production, dissemination and negotiation of dedicated artefacts. The artefacts may have different shapes: They may come as interactive visualizations, open datasets or open-source computer models. Nevertheless, they have conceptual similarities. All artefacts produced should potentially be accessible and usable by others without the need to consult the researchers who produced them. The operationalization of this objective strongly preferentiates certain characteristics: Artefacts should be digital, mobile, documented, addressable and accessible across the networked infrastructures of the internet. Put differently, openness is equated with *digital openness.* Referring to historian Theodore Porter (1996), this can be characterized as a strategy of impersonality in science favoring explicit quantified information over embodied expert knowledge. Building on his historical study of accounting practices in the 19th and 20th century, Porter has shown how quantification and statistics emerged as technologies of trust constructing new kinds of legitimacy for economic and political actors through distancing and objectivation (ibid.: 87ff). Echoing Porter, Matthias Heymann, Gabriele Gramelsberger and Martin Mahony have argued that computer modeling and simulation have to be understood as a more recent toolbox of objectivation to exercise epistemic and political power. Their most consequential promise is to provide an objective means to make claims about the future (Heymann et al. 2017). In the present book, I show how the practices, artefacts and digital openness[138] have to be considered as another push towards the depersonalization of scientific knowledge. Of course, science has always worked at making scientific knowledge independent from the researcher and to translate it into a standardized written form that can be shared within communities. The academic paper is the ultimate example of such an immutable mobile for scientific knowledge (Latour 1988). The mobility of the scientific paper, however, appears to be highly limited. Typically, such artefacts only travel between fellow researchers of the same scientific community that shares a common esoteric language. They rarely move across disciplinary borders and beyond the scientific community. For such translation of scientific knowledge across social worlds, the researcher represents a preferential

138 I became aware just before publishing this book that Maximilian Heimstädt had already developed scientific contributions characterizing the term of *digital openness* (e.g. Heimstädt 2014). For production-related timelines, it was not possible to discuss this valuable content further in this text.

passage point to go through. Researchers, therefore, maintain considerable control over the dissemination of the scientific knowledge produced in their own research projects. In the context of environmental science, the PIK and its long-time director John Schellnhuber have perfected the role of scientists as passage points for future knowledge on climate change. Empowered through computer technology, they act as exclusive (human) translators of the phenomena and trends described within their simulations. Within this function, they are consulted by politicians, technocrats, teachers, journalists and company leaders. The flip side of the coin is that they frequently become the victims of 'shoot the messenger' maneuvers by climate skeptics and deniers. A shift of power is taking place in the course of the reconfiguration of scientific practice towards digital openness. Scientific knowledge is not primarily embodied by the researcher nor by artefacts (publications, scientific models). Instead, it is embodied by distributed, networked infrastructures, which often comprise scientific, commercial and media-public spheres. The scientist is one of the nodes of these novel actor-networks, but only one of them. Equally important are novel elements that enable forms of stabilization in increasingly fluid environments. It is not only a matter of standardizing data and programming code but also of a situated publication of the constituted knowledge within certain infrastructures and communities. Depending on the analytical perspective, these elements can be described as *anchoring devices* of knowledge organization or as *viscous elements* within the fluid socio-technical networks of digital technology. Constructing and managing both mobile artefacts and anchoring devices entails strong characteristics of futurework.

New kinds of futurework

Climate impact researchers are accustomed to taking the future into account. Their daily work involves practices of future imagination, the formalization of these imaginations in computer models and simulations, and the representation of insights from such formalized prophecy in worlds outside science. The reconfigurations of scientific practice described in this book, however, add novel features to climate impact research as futurework. The new 'Scientists for Future' should not only make scientifically sound projections of the risky developments ahead. They are also expected to empower others to make such claims about the future. Everything – data, models, visualizations, guidelines – should be made prospectively accessible, comprehensible and reusable for others as early as possible. The prime example of an operationalization of this practice is the work around CLIMADA (Chapter IV and V). Here, models, datasets and documentations are optimized to help others, namely, risk modelers in the insurance industry, to take climate change into account within their own calculation work.

These new configurations in modeling practice hailing permeability and digital openness are illustrated in Figure 57. The vertical blue lines and bullets designate the multiple digital artefacts produced, mediated and disseminated throughout the modeling process. The thinner lines represent the embedding of the artefacts within digital, distributed and networked infrastructures and their blue arrows the prospective outlook for future use.

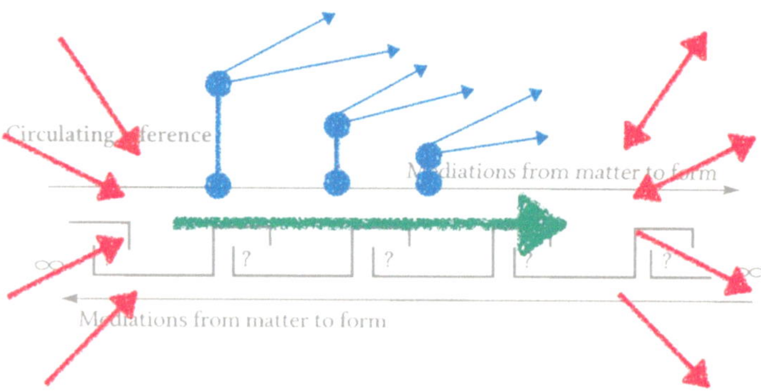

Figure 57: Permeability and digital openness.
Source: Latour 1999a supplemented by additional elements

Investigating achievements and frictions around digital openness

On the one hand, by investing in permeability and digital openness, climate impact modelers influence the way modelers from other fields (reinsurance companies, banks, state authorities) calculate the future. They have their say regarding what scenarios are computationally imaginable. On the other hand, this new kind of futurework takes time and resources. Researchers spend fewer of their working hours thinking about the inner logic and performance of their simulations and more designing explainable and interpretable tools and resources for future use. More than before, scientists have to acquire new professional skills and become veritable designers producing digital artefacts for future use by prospective audiences.

Similar to other major disruptions of scientific practice, the transformation of modeling towards permeability and digital openness does not always happen smoothly and often creates frictions between expectations and actual practice. I have addressed such frictions within Chapter III to VI. While online platforms providing datasets, mobile models and visualizations operate with strategies of impersonality, they tend to obscure the extensive translation work undertaken by multiple human mediators. Neither open data, models nor visualizations should be taken as given; they have to be explained, curated and made relevant by dedicated people. Despite the promises of digital openness to make knowledge accessible everywhere for anyone at all times, knowledge-storing artefacts have to be constantly maintained and actively mediated in order to remain open. I have tried to value this invisible work enabling and powering digital openness in science throughout this book. As such, I have highlighted the work of science communicators, data storytellers and scientific programmers, as well as maintenance workers of data centers. Moreover, I also argued that digital openness can be interpreted as a transitional phase preparing the age of linked data and learning machines. Different than official discourse suggests, digital openness is not primarily fit to make knowledge accessible for humans but for machines. Agency may, therefore, gradually shift to new kinds of collectives (intelligent systems) gathering human (data scientists) and nonhuman actors (algorithms, data, digital infrastructure). While this development is currently only in its infancy, digital openness can be seen as the future imaginary preparing this new generation of cybernetic practice.

Acknowledgements

Many people have provided intellectual input, settings of debate and organizational help to make this book possible. Among them are (in alphabetical order) Pablo Abend, Ines Blumenthal, Asher Boersma, Carolin Gerlitz, Robert Gieseke, Sebastian Giessmann, Gabriele Gramelsberger, Paul Heinicker, Claudia Müller-Birn, Asko Lehmuskallio, Thomas Nocke, Anna Lisa Ramella, Philip Saunders, Birgit Schneider, Cornelius Schubert, Erhard Schüttpelz, Tristan Thielmann, all my interview partners at PIK and elsewhere, and my family. I am sincerely grateful to all of you.

Literature

Abend, Pablo (2018). "Seasons change, so do we: heterogeneous temporalities, algorithmic frames and subjective time in geomedia", in: Lammes, S., Perkins, C., Gekker, A., Hind, S., Wilmott, C., Evans, D. (Eds.), *Time for Mapping: Cartographic Temporalities.* Manchester University Press.

Abend, Pablo (2013): *Geobrowsing.* Bielefeld.

Abend, Pablo et al. (2012): "Geobrowsing Behaviour in Google Earth – A Semantic Video Content Analysis of On-Screen Navigation", in: *GI_Forum* 12.

Ackoff, Russell L. (1989): "From Data to Wisdom", in: *Journal of Applied Systems Analysis* 16 (1), 3-9.

Adler, Mortimer Jerome (1986): A Guidebook to Learning: For a Lifelong Pursuit of Wisdom, London.

AI Now (2018): *2018 Report.* New York.

Anand, Nikhil et al. (Eds.) (2018): *The Promise of Infrastructure*, Durham.

Asayama, Shinichiro et al. (2019): "Beyond Solutionist Science for the Anthropocene: To Navigate the Contentious Atmosphere of Solar Geoengineering", in: *The Anthropocene Review* 6 (1-2), 19-37.

Bailey, Diane E. et al. (2012): "The Lure of the Virtual", in: *Organization Science* 23 (5), 1485-1504.

Baker, Monya (2016): "1,500 Scientists Lift the Lid on Reproducibility", in: *Nature* 533 (7604), 452-454.

Barnes, Jessica/Dove, Michael R. (2015): *Climate Cultures: Anthropological Perspectives on Climate Change*, Yale University Press.

Barry, Andrew et al. (2008): "Logics of Interdisciplinarity", in: *Economy and Society* 37 (1), 20-49.

Bateson, Gregory (1987): Steps to an Ecology of Mind: Collected Essays in Anthropology, Psychiatry, Evolution, and Epistemology, Northvale, N.J.

Beck, Ulrich (2002): "The Terrorist Threat: World Risk Society Revisited", in: *Theory, Culture & Society* 19 (4), 39-55.

Bender, Cora/Zillinger, Martin (2015): *Handbuch der Medienethnographie*, 1st edition, Berlin.

Berners-Lee, Tim (2009): "Linked Data – Design Issues", www.w3.org /DesignIssues /LinkedData.html, 08.08.2019.

Blumenthal, Ines et al. (2016a): Vorschläge für die Vermittlung des Klimawandels im Bildungskontext, Potsdam.

Blumenthal, Ines et al. (2016b): "Climate Impacts for German Schools – an Educational Web Portal Solution", in *Handbook of Climate Change Communication*, New York.

Boch, Ralph (2008): Der Potsdamer Telegrafenberg: ein traditionsreicher Forschungsstandort zwischen DDR und wiedervereinigtem Deutschland; zum Festakt anlässlich der Verabschiedung von Herrn Prof. Dr. Dr. h.c. Rolf Emmermann am 8.1.2008, Potsdam, http://gfz-public.gfz-potsdam.de/pubman/item/escidoc:8736, 16.07.2019

Bolter, J. David/Gromala, Diane (2003): Windows and Mirrors: Interaction Design, Digital Art, and the Myth of Transparency, Cambridge, Mass.

Borra, Erik/Rieder, Bernhard (2014): "Programmed method: developing a toolset for capturing and analyzing tweets", *Aslib Journal of Information Management* 66 (3), 262-278.

Bowker, Geoffrey C. (1994): "Information Mythology: The World of/as Information", in: Lisa Bud-Frierman (Ed.), *Information Acumen: The Understanding and Use of Knowledge in Modern Business*, Andover, UK, 231-247.

Bowker, Geoffrey C. (2005): Memory Practices in the Sciences. Inside Technology, Cambridge, Mass.

Bowker, Geoffrey C. (2015): "Temporality", Cultural Anthropology Website, 24.09.2015, https://culanth.org/fieldsights/temporality, 02.04.2019.

Bredekamp, Horst, et al. (Eds.) (2012): Das Technische Bild: Kompendium zu einer Stilgeschichte wissenschaftlicher Bilder, Berlin/Boston.

Bresch, David (2015): "CLIMADA Manual", Software manual, 23.01.2017, https://climate-adapt.eea.europa.eu/metadata/tools/climada, 02.04.2019.

Bundesregierung (2018): "Nationale KI-Strategie", Nationale Strategie für Künstliche Intelligenz, https://www.ki-strategie-deutschland.de/home.html, 02.04.2019.

Burkhardt, Marcus (2015): Digitale Datenbanken. Eine Medientheorie Im Zeitalter von Big Data, Bielefeld.

Callon, Michel (1984): "Some Elements of a Sociology of Translation: Domestication of the Scallops and the Fishermen of St Brieuc Bay", in: *The Sociological Review* 32 (S1), 196-233.

Carlyle, Thomas (1993): *On Heroes, Hero-Worship, and the Heroic in History*, Vol. 1, Berkeley, Calif.

Charney, Jule et al. (1979). Carbon Dioxide and Climate: A Scientific Assessment (Ad Hoc Study Group on Carbon Dioxide and Climate). National Academy of Sciences, Washington D.C.

Clarke, Adele E. (2003): "Situational Analysis. Grounded Theory Mapping after the Postmodern Turn", 2003.

Computing Research Association (2016): "Computing Research and the Emerging Field of Data Science", in: *CRA Bulletin* (blog), 07.10.2016, https://cra.org/data-science/.

Costello, Mark J. (2009): "Motivating Online Publication of Data", in: *BioScience* 59 (5), 418-427.

Crampton, Jeremy W. et al. (2013): "Beyond the geotag: situating 'big data' and leveraging the potential of the geoweb." in: *Cartography and Geographic Information Science* 40, 130–139. https://doi.org/10.1080/15230406.2013.777137

Crawford, Kate (2013): “The Hidden Biases in Big Data”, in: *Harvard Business Review,* 01.04.2013, https://hbr.org/2013/04/the-hidden-biases-in-big-data, 02.04.2019.

Crawford, Kate/Calo, Ryan (2016): “There Is a Blind Spot in AI Research”, *Nature* 538 (7625), 311-313.

Creative Commons (2019): “CC BY 4.0.” 2019, https://creativecommons.org/licenses/by/4.0/.

Crutzen, Paul J. (2006): “The ‘Anthropocene’”, in: Eckart Ehlers/Thomas Krafft (Eds.), *Earth System Science in the Anthropocene,* Berlin/Heidelberg, 13-18.

De Laet, Marianne/Mol, Annemarie (2000): “The Zimbabwe Bush Pump: Mechanics of a Fluid Technology”, in: *Social Studies of Science* 30 (2), 225-263.

Derrida, Jacques (1995): *Of Grammatology,* Baltimore.

Dessai, Suraje/Hulme, Mike (2007): “Assessing the Robustness of Adaptation Decisions to Climate Change Uncertainties: A Case Study on Water Resources Management in the East of England”, in: *Global Environmental Change* 17 (1), 59-72.

Dodge, Martin et al. (2011): Rethinking Maps: New Frontiers in Cartographic Theory, 1st edition, London.

Donges, Jonathan F. et al. (2017): “Closing the Loop: Reconnecting Human Dynamics to Earth System Science”, in: *The Anthropocene Review* 4 (2), 151-157.

Downey, Allen (2012): *Think Python,* Sebastopol, Calif.

Doyle, Julie (2009): “Seeing the Climate? The Problematic Status of Visual Evidence in Climate Change Campaigning”, in: Sidney Dobrin/Sean Morey (Eds.), *Ecosee: Image, Rhetoric, Nature,* New York, 279-298.

Doyle, Julie (2011): *Mediating Climate Change,* Burlington, VT.

Edwards, Paul N. (2001): “Representing the Global Atmosphere”, in: Clark A. Miller/Paul N. Edwards (Eds.), *Changing the Atmosphere. Expert Knowledge and Environmental Governance,* Cambridge, Mass.

Edwards, Paul N. (2010): A Vast Machine: Computer Models, Climate Data, and the Politics of Global Warming, Cambridge, Mass.

Engeström, Yrjö (1990): "When Is A Tool?", in: Learning, Working and Imagining, Twelve Studies in Activity Theory, Helsinki.

Falzon, Mark-Anthony (Ed.) (2016): Multi-Sited Ethnography: Theory, Praxis and Locality in Contemporary Research, London.

Faraj, Samer et al. (2011): "Knowledge Collaboration in Online Communities." *Organization Science* 22, 1224–1239. https://doi.org/10.1287/orsc.1100.0614

Fecher, Benedikt/Friesike, Sascha (2014): "Open Science: One Term, Five Schools of Thought", in: Sönke Bartling/Sascha Friesike (Eds.), *Opening Science*, Cham, 17-47.

Fine, Gary Alan (1979): "Small Groups and Culture Creation: The Idioculture of Little League Baseball Teams", in: *American Sociological Review* 44 (5), 733.

Fine, Gary Alan (2002): Shared Fantasy: Role-Playing Games as Social Worlds, Reprint, Chicago, Ill.

Fine, Gary Alan (2007): Authors of the Storm: Meteorologists and the Culture of Prediction, Chicago, Ill./Bristol.

Fine, Gary Alan (2008): Kitchens: The Culture of Restaurant Work by Gary Alan Fine, Berkeley, Calif.

Floridi, Luciano (2010): *Information: A Very Short Introduction*, 1st edition, Oxford/New York.

Frické, Martin (2009): "The Knowledge Pyramid: A Critique of the DIKW Hierarchy", in: *Journal of Information Science* 35 (2), 131-142.

Füssel, Hans M. (2005): "Vulnerability in Climate Change Research: A Comprehensive Conceptual Framework", UC Berkeley: University of California International and Area Studies, https://escholarship.org/uc/item/8993z6nm.

Galle, A. (1926): "Zur Geschichte des Potsdamer Telegrafenberges", in: *Das Weltall* 5, 65.

Geiger, Tobias (2018): "Continuous National Gross Domestic Product (GDP) Time Series for 195 Countries: Past Observations (1850–2005)

Harmonized with Future Projections According to the Shared Socio-Economic Pathways (2006–2100)", in: *Earth System Science Data* 10 (2), 847-856.

Geiger, Tobias et al. (2017): "Spatially-Explicit Gross Cell Product (GCP) Time Series: Past Observations (1850-2000) Harmonized with Future Projections According to the Shared Socioeconomic Pathways (2010-2100)", GFZ Data Services, https://doi.org/10.5880/pik.2017.007.

Gerstengarbe, Friedrich-Willhelm et al. (Eds.) (2013): Zwei Grad mehr in Deutschland: Wie der Klimawandel unseren Alltag verändern wird, Frankfurt a.M.

GFZ, Helmholtz-Zentrum Potsdam – Deutsches GeoForschungsZentrum (2012): "Und der Pendelsaal – die Schatzkammer des T-Bergs ist fertig", in: *GeoForschungsZeitung*, April 2012, http://gfzpublic.gfz-potsdam.de/pubman/item/escidoc:65071/component/escidoc:65070/GFZeitung_2012_04_10.pdf.

GFZ, Helmholtz-Zentrum Potsdam – Deutsches GeoForschungsZentrum (2017): *Fokus: Erde: Focus: Earth*, Berlin.

Gieryn, Thomas F. (1983): "Boundary-Work and the Demarcation of Science from Non-Science: Strains and Interests in Professional Ideologies of Scientists", in: *American Sociological Review* 48 (6), 781.

Gieryn, Thomas F. (2002): "What Buildings Do", in: *Theory & Society* 31 (1), 35-74.

Gieseke, Robert et al. (2018): "Pymagicc: A Python Wrapper for the Simple Climate Model MAGICC", in: *The Journal of Open Source Software* 3 (22), 516.

Giorgi, F. (2008): "WMO Bulletin", *WMO Bulletin*, 2008, https://public.wmo.int/en/resources/bulletin, 04.09.2019.

Girod, Bastien et al. (2009): "The Evolution of the IPCC's Emissions Scenarios", in: *Environmental Science & Policy* 12 (2) 103-118.

Gitelman, Lisa (2013):*"Raw Data" Is an Oxymoron*, Cambridge, Mass.

Gitelman, Lisa/Jackson, Virginia (2013): "Introduction", in: *"Raw Data" Is an Oxymoron*, Cambridge, Mass., 1–14.

Glaser, Barney/Strauss, Anselm (1999): The Discovery of Grounded Theory: Strategies for Qualitative Research, Chicago, Ill.

Goldblatt Colin/Watson, Andrew J. (2012): "The Runaway Greenhouse: Implications for Future Climate Change, Geoengineering and Planetary Atmospheres", in: *Philosophical Transactions of the Royal Society A: Mathematical, Physical and Engineering Sciences* 370 (1974), 4197-4216.

Goodwin, Charles (1994): "Professional Vision", in: *American Anthropologist*, 96 (3), 606-633.

Gore, Al (1998): "The Digital Earth: Understanding Our Planet in the 21st Century", in: *Australian Surveyor* 43 (2), 89-91.

Gramelsberger, Gabriele (2008a): Computerexperimente: Zum Wandel der Wissenschaft im Zeitalter des Computers, Bielefeld.

Gramelsberger, Gabriele (2008b): "Das Epistemische Gewebe Simulierter Welten", in: Andrea Gleiniger et al. (Eds.), *Simulation: Praesentationstechnik Und Erkenntnisinstrument*, Basel, 83–91.

Gramelsberger, Gabriele/Feichter, Johann (Eds.) (2011): *Climate Change and Policy*, Berlin/Heidelberg.

Gupta, Akhil/Ferguson, James (1997): "Discipline and Practice: 'The Field' As Site, Method, and Location in Anthropology", in: Akhil Gupta/James Ferguson (Eds.), *Anthropological Locations: Boundaries and Grounds of a Field Science*, Berkeley, Calif.

Gurstein, Michael B. (2011): "Open Data: Empowering the Empowered or Effective Data Use for Everyone?", in: *First Monday* 16 (2). https://firstmonday.org/article/view/3316/2764

Gurstein, Michael B. (2013): "Should 'Open Government Data' Be a Product or a Service (and Why Does It Matter?)", in: *Gurstein's Community Informatics*, 03.02.2013. https://gurstein.wordpress.com/2013/02/03/is-open-government-data-a-product-or-a-service-and-why-does-it-matter/, 02.04.2019.

Hagedorn, G. et al. (2019): "The Concerns of the Young Protesters Are Justified. A Statement by Scientists for Future Concerning the Protests for More Climate Protection", in: *GAIA* 28 (2), 79-87.

Haraway, Donna (1988): “Situated Knowledges: The Science Question in Feminism and the Privilege of Partial Perspective”, in: *Feminist Studies* 14 (3), 575.

Harley, J. Brian (1988): “Maps, Knowledge, and Power”, in: Denis Cosgrove/Stephen Daniels (Eds.), *The Iconography of Landscape: Essays on the Symbolic Representation, Design and Use of Past Environments*, Avon, 277-312.

Harper, Douglas (2002): “Talking about Pictures: A Case for Photo Elicitation”, in: *Visual Studies* 17 (1), 13-26.

Hastrup, Kirsten/Skrydstrup, Martin (2014): *The social life of climate change models: anticipating nature.* Routledge, New York; London.

Heberger, Matthew et al. (2009): *The Impacts of Sea-Level Rise on the California Coast*, San Francisco, Calif.

Heidegger, Martin (1954): “Die Frage nach der Technik ”, in: *Vorträge und Aufsätze* 8, 13–55.

Heimstädt, Maximilian (2014): “The Institutionalization of Digital Openness: How NGOs, Hackers and Civil Servants Organize Municipal Open Data Ecosystems”, in: *Proceedings of The International Symposium on Open Collaboration.* Presented at the OpenSym ’14: The International Symposium on Open Collaboration, ACM, Berlin Germany. https://doi.org/10.1145/2641580.2641626.

Henderson, Kathryn (1991): “Flexible Sketches and Inflexible Data Bases: Visual Communication, Conscription Devices, and Boundary Objects in Design Engineering”, in: *Science, Technology & Human Values* 16 (4), 448-473.

Herold, David K. et al. (2013): Understanding Creative Users of ICTs: Users as Social Actors. Routledge, New York; London.

Herrmann, D. B. (1975): “Zur Vorgeschichte des Astrophysikalischen Observatorium Potsdam (1865 bis 1874)”, in: *Astronomische Nachrichten* 296 (6), 245-259.

Hewitson, Bruce et al. (2017): “Climate Information Websites: An Evolving Landscape: Climate Information Websites”, in: *Wiley Interdisciplinary Reviews: Climate Change* 8 (5), e470.

Heymann, Matthias et al. (2017): "Key Characteristics of Cultures of Prediction", in: Matthias Heymann et al. (Eds.), Cultures of Prediction in Atmospheric and Climate Science: Epistemic and Cultural Shifts in Computer-Based Modelling and Simulation, 1st edition, London, 18-41.

Heymann, Matthias/Hundebøl, Nils R. (2017): "From Heuristic to Predictive. Making Climate Models into Political Instruments", in: Matthias Heymann et al. (Eds.), *Cultures of Prediction in Atmospheric and Climate Science: Epistemic and Cultural Shifts in Computer-Based Modelling and Simulation*, 1st edition, London, 100–119.

Hirschauer, Stefan/Amann, Klaus (1997): Die Befremdung der eigenen Kultur: zur ethnographischen Herausforderung soziologischer Empirie, Frankfurt a.M.

Hoffmann, Anne (2017): "25 Jahre Potsdam-Institut Für Klimafolgenforschung – Damals hat man das Institut fast für einen Scherz gehalten", RBB24 Rundfunk Berlin-Brandenburg (Online), 10.11.2017, https://www.rbb24.de/panorama/beitrag/2017/10/potsdam-klimafo, 02.04.2019.

Hughes, Thomas Parker (1983): Networks of Power: Electrification in Western Society, 1880-1930, Revised, Baltimore, Md.

Hullmann, Angela (2008): European Activities in the Field of Ethical, Legal and Social Aspects (ELSA) and Governance of Nanotechnology, DG Research, Brussels.

Hulme, Mike (2008): "Geographical Work at the Boundaries of Climate Change: Boundary Crossings", in: *Transactions of the Institute of British Geographers* 33 (1), 5-11.

Hulme, Mike (2009): Why We Disagree about Climate Change: Understanding Controversy, Inaction and Opportunity, 4th edition, Cambridge, UK/New York.

Hulme, Mike (2011): "Reducing the Future to Climate: A Story of Climate Determinism and Reductionism", in: *Osiris* 26 (1), 245-266.

Hulme, Mike (2016): *Weathered: Cultures of Climate*, Thousand Oaks, Calif.

Hulme, Mike/Mahony, Martin (2010): "Climate Change: What Do We Know about the IPCC?", in: *Progress in Physical Geography: Earth and Environment* 34 (5), 705-718.

Hurtig, E. (Ed.) (1890): Die königlichen Observatorien für Astrophysik, Meterologie und Geodäsie bei Potsdam, herausgegeben von den betheiligten Directoren, Berlin.

International Telecommunication Union (2018): United Nations Activities on Artificial Intelligence (AI), Geneva.

IPCC: Intergovernmental Panel on Climate Change (2000): Emissions Scenarios. A Special Report of IPCC Working Group III. Summary for Policy Makers, Geneva.

IPCC: Intergovernmental Panel on Climate Change (2013): 5th Assessment Report: The Physical Science Base (Summary for Policy Makers), Geneva.

Jacob, Daniela et al. (2017): „Regionale Klimamodellierung", in: *Klimawandel in Deutschland.* Springer Spektrum, Berlin, Heidelberg, pp. 27–35.

Jasanoff, Sheila (2010): "A New Climate for Society", in: *Theory, Culture & Society* 27 (2-3), 233-253.

Jasanoff, Sheila/Kim, Sang-Hyun (2015): Dreamscapes of Modernity: Sociotechnical Imaginaries and the Fabrication of Power, Chicago, Ill.

Jensen, Michael C./Meckling, William H. (1976): "Theory of the Firm: Managerial Behavior, Agency Costs and Ownership Structure", in: *Journal of Financial Economics* 3 (2), 305-360.

Kalthoff, Herbert (2010): "Beobachtung und Komplexität", in: *Sozialer Sinn: Zeitschrift für hermeneutische Sozialforschung* 11 (2).

Karpatne, Anuj et al. (2017): "Machine Learning for the Geosciences: Challenges and Opportunities", in: *ArXiv:1711.04708 [Physics]*, November.

Kasting, James F. et al. (1988): "How Climate Evolved on the Terrestrial Planets", in: *Scientific American* 258 (2), 90-97.

Kautzleben, Heinz (1999): "Das Zentralinstitut für Physik der Erde und die in ihm aufgegangenen Institute der Deutschen Akademie der

Wissenschaften zu Berlin im Zeitraum von 1950 bis 1973", in: *Die Berliner Akademie von 1950 bis 1972. Sitzungsberichte der Leibniz-Sozietät*, 22.

Kitchin, Rob (2014a): "Big Data, new epistemologies and paradigm shifts", in: *Big Data & Society* 1, 1-12.

Kitchin, Rob (2014b): The Data Revolution: Big Data, Open Data, Data Infrastructures and Their Consequences, Thousand Oaks, Calif.

Kitchin, Rob et al. (2013): "Unfolding Mapping Practices: A New Epistemology for Cartography", in: *Transactions of the Institute of British Geographers* 38 (3), 480-496.

Kitchin, Rob/Lauriault, Tracey (2014): "Towards Critical Data Studies: Charting and Unpacking Data Assemblages and Their Work", The Programmable City Working Paper 2; pre-print version of chapter to be published in: J. Eckert et al. (Eds.), *Geoweb and Big Data*, Lincoln, NE.

Kluyver, Thomas et al. (2016): *Jupyter Notebooks –a publishing format for reproducible computational workflows*. https://doi.org/10.3233/978-1-61499-649-1-87

Knorr-Cetina, Karin (2003): *Epistemic Cultures: How the Sciences Make Knowledge*, Cambridge, Mass./London.

Knorr Cetina, Karin (2009): "The Synthetic Situation: Interactionism for a Global World", in: *Symbolic Interaction* 32 (1), 61-87.

Kouw, Matthijs (2010): "The Craft of Modeling, the Modeling of Craft", Presentation at EASST 2010 Conference, www.homepages.ucl.ac.uk/~ucessjb/EASST%202010/Kouw%20EASST%202010.pdf, 02.04.2019.

Kratz, John/Strasser, Carly (2014): "Data publication consensus and controversies", in: *F1000Research* 3, 94.

Krauss, Werner/von Storch, Hans (2012): "Post-Normal Practices between Regional Climate Services and Local Knowledge", in: *Nature and Culture* 7 (2).

Krishnan, Armin (2009): "What Are Academic Disciplines? Some Observations on the Disciplinarity vs. Interdisciplinarity Debate", in: *NCRM Working Paper Series* 9 (3).

Kühnen, Friedrich/Furtwängler, Philip (1906): Bestimmung der absoluten Grösze der Schwerkraft zu Potsdam: Mit Reversionspendeln, Berlin.

Kvale, Steinar (2007): "Interview Variations", in: Steinar Kvale (Ed.), *Doing Interviews*, London.

Lahsen, Myanna (2005): "Seductive Simulations? Uncertainty Distribution around Climate Models", in: *Social Studies of Science* 35 (6), 895-922.

Lammes, Sybille et al. (Eds.) (2018): *Time for Mapping: Cartographic Temporalities*, 1st edition, Manchester.

Latour, Bruno (1987): Science in Action: How to Follow Scientists and Engineers through Society, Cambridge, Mass.

Latour, Bruno (1988): "Visualisation and Cognition: Drawing Things Together", in: *Knowledge and Society Studies in the Sociology of Culture Past and Present* 6, 1-40.

Latour, Bruno (1999a): "Circulating Reference", in: *Pandora's Hope: Essays on the Reality of Science Studies*, 1st edition, Cambridge, Mass., 24–79.

Latour, Bruno (1999b): Pandora's Hope: Essays on the Reality of Science Studies, 1st edition, Cambridge, Mass.

Lave, Jean (1991): "Situating Learning in Communities of Practice", in: *Perspectives on Socially Shared Cognition* 2, 63-82.

Law, John (2007): "Making a Mess with Method", in: William Outhwaite/Stephen P. Turner (Eds.), *The Sage Handbook of Social Science Methodology*, London, 595-606..

Leonardi, Paul M. (2012): Car Crashes without Cars: Lessons about Simulation Technology and Organizational Change from Automotive Design, Cambridge, Mass.

Leonelli, Sabina (2015): "What Counts as Scientific Data? A Relational Framework", in: *Philosophy of Science* 82 (5), 810-821.

Luhmann, Niklas (2014): Vertrauen: Ein Mechanismus der Reduktion sozialer Komplexität, 5th edition, Konstanz.

Lury, Celia/Wakeford, Nina (2014): "Introduction", in: Celia Lury/Nina Wakeford (Eds.), *Inventive Methods: The Happening of the Social*, Oxford, UK, 1-24.

MacKenzie, Adrian (2006): Cutting Code: Software and Sociality, Bern.

MacKenzie, Donald/Spinardi, Graham (1995): "Tacit Knowledge, Weapons Design, and the Uninvention of Nuclear Weapons", in: *American Journal of Sociology* 101 (1), 44-99.

Mahoney, James (2000) "Path Dependence in Historical Sociology", in: *Theory and Society* 29 (4), 507-548.

Mahony, Martin (2017): "The (Re)Emergence of Regional Climate: Mobile Models, Regional Visions and the Government of Climate Change", in: Matthias Heymann et al. (Eds.), *Cultures of Prediction in Atmospheric and Climate Science: Epistemic and Cultural Shifts in Computer-Based Modelling and Simulation*, 1st edition, London, 139-158.

Mahony, Martin/Hulme, Mike (2012): "Model Migrations: Mobility and Boundary Crossings in Regional Climate Prediction", in: *Transactions of the Institute of British Geographers* 37 (2), 197-211.

Mahony, Martin/Hulme, Mike (2014): "The Color of Risk: Expert Judgement and Diagrammatic Reasoning in the IPCC's 'Burning Embers'", in: Birgit Schneider/Thomas Nocke (Eds.), *Image Politics of Climate Change. Visualizations, Imaginations, Documentations*, Bielefeld, 105-124.

Manzo, Kate (2010): "Beyond Polar Bears? Re-Envisioning Climate Change", in: *Meteorological Applications* 17 (2), 196-208.

Marcus, George E. (1995): "Ethnography in/of the World System: The Emergence of Multi-Sited Ethnography", in: *Annual Review of Anthropology* 24 (1), 95-117.

Marres, Noortje (2017): Digital Sociology: The Reinvention of Social Research, Malden, MA.

Marres, Noortje/Gerlitz, Carolin (2015): "Interface Methods: Renegotiating Relations between Digital Social Research, STS and Sociology", in: *Sociological Review* 64 (1), 21-46.

Marwick, Ben (2015): "Reproducibility: How Computers Broke Science – and What We Can Do to Fix It", in: *The Conversation*. https://theconversation.com/how-computers-broke-science-and-what-we-can-do-to-fix-it-49938, 03.09.2019.

McArthur, Victoria (2009): "Communication technologies and cultural identity a critical discussion of ICTs for development", in: *2009 IEEE Toronto International Conference Science and Technology for Humanity (TIC-STH)*, Toronto, ON. https://doi.org/10.1109/TIC-STH.2009.5444367

McCarthy, James J. et al. (2001): *Climate change 2001: Impacts, adaptation, and vulnerability: contribution of Working Group II to the third assessment report of the Intergovernmental Panel on Climate Change.* Cambridge University Press.

Meinshausen, Malte at al. (2011): "Emulating Coupled Atmosphere-Ocean and Carbon Cycle Models with a Simpler Model, MAGICC6 – Part 1: Model Description and Calibration", in: *Atmospheric Chemistry and Physics* 11 (4), 1417-1456.

Michelson, Albert A. (1881): *The Relative Motion of the Earth and the Luminiferous Ether*, https://en.wikisource.org/wiki/The_Relative_Motion_of_the_Earth_and_the_Luminiferous_Ether, 02.04.2019.

Mitchell, William J. T. (1987): *Iconology: Image, Text, Ideology*, 2nd edition, Chicago, Ill.

Moser, Susanne C./Dilling, Lisa (2007): Creating a Climate for Change: Communicating Climate Change and Facilitating Social Change, Cambridge, UK/New York.

Moss, Richard H. et al. (2010): "The Next Generation of Scenarios for Climate Change Research and Assessment", in: *Nature* 463 (7282), 747-756.

Müller, Marion G. (2011): "Iconography and Iconology as a Visual Method and Approach", in: Luc Pauwels/Dawn Manny (Eds.), *The SAGE Handbook of Visual Research Methods*, Thousand Oaks, Calif. et al., 283-297.

Neset, Tina-Simone et al. (2016): "Map-Based Web Tools Supporting Climate Change Adaptation", in: *The Professional Geographer* 68 (1), 103-114.

Neverla, Irene (2012): Das Medien-Klima. Fragen und Befunde der kommunikationswissenschaftlichen Klimaforschung, Wiesbaden.

Nicholson-Cole, Sophie A. (2005): "Representing Climate Change Futures: A Critique on the Use of Images for Visual Communication", in: *Computers, Environment and Urban Systems* 29 (3), 255-273.

National Oceanic and Atmospheric Administration (NOAA) Coastal Services Center (2012): "Lidar 101: An Introduction to Lidar Technology, Data, and Applications." Revised. Charleston, SC.

Norris, Pippa (2001): Digital Divide: Civic Engagement, Information Poverty, and the Internet Worldwide, Cambridge UK.

Odendahl, Teresa/Shaw, Aileen M. (2001): "Interviewing Elites", in: Jaber F. Gubrium/James A. Holstein (Eds.), *Handbook of Interview Research*, Thousand Oaks, Calif., 299-316.

O'Neil, Cathy/Schutt, Rachel (2013) *Doing Data Science: Straight Talk from the Frontline*, 1st edition, Beijing/Sebastopol.

Orlowski, Boris (2007): *STARS*, Dissertation, University of Hamburg.

Oudshoorn, Nelly/Pinch, Trevor (2005): How Users Matter: The Co-Construction of Users and Technology, Cambridge, Mass.

Panofsky, Erwin (1972): Studies in Iconology. Humanistic Themes in the Art of the Renaissance, Boulder, CO.

Perkins, Chris (2014): "Plotting practices and politics: (im)mutable narratives in OpenStreetMap", *Transactions of the Institute of British Geographers* 39, 304–317. https://doi.org/10.1111/tran.12022

Pias, Claus (2008): "Klimasimulation", in: Petra Lutz/Thomas Macho (Eds.), *2°: das Wetter, der Mensch und sein Klima*, Göttingen, 108-119.

Pickering, Andrew (2002): "Cybernetics and the Mangle: Ashby, Beer and Pask", in: *Social Studies of Science* 32 (3), 413-437.

Pierce, Roger (2008): "Asking Questions: Effective Elite Interviews, Other Interviews, Vignettes, Projective Questions, and Focus Groups," in: *Research Methods in Politics*, London.

PIK (2012): "Impacts of Climate Change. Internet Portal Climate Impacts Online", brochure, Potsdam.

Rammert, Werner (2007): "Technografie trifft Theorie. Forschungsperspektiven einer Soziologie der Technik", in: *TUTS – Working Papers* 1-2007, 37.

Rammert, Werner/Schubert, Cornelius (2006): *Technografie: zur Mikrosoziologie der Technik*, Frankfurt a.M.

Reicheneder, Karl (1959): "Reference-Value of Gravity at Potsdam", in: *Bulletin Géodésique (1946-1975)* 51 (1), 80-81.

Rheinberger, Hans-Jörg (1997): Toward a History of Epistemic Things: Synthesizing Proteins in the Test Tube, 1st edition, Stanford, Calif.

Rheinberger, Hans-Jörg (2011): "Infra-Experimentality: From Traces to Data, from Data to Patterning Facts", in: *History of Science* 49 (3), 337-348.

Ribes, David (2018): "STS, Meet Data Science, Once Again", in: *Science, Technology, & Human Values* September, 016224391879889.

Rieder, Bernhard (2015): *YouTube Data Tools* (Version 1.22) [Software]. https://tools.digitalmethods.net/netvizz/youtube/.

Rockström, Johan et al. (2009): "A Safe Operating Space for Humanity", in: *Nature* 461 (7263), 472-475.

Rogers, Richard (2013): *Digital Methods*, Cambridge, Mass.

Rogers, Richard (2015): "Digital Methods for Web Research", in: Robert A Scott/Stephen M. Kosslyn (Eds.), *Emerging Trends in the Social and Behavioral Sciences: An Interdisciplinary, Searchable, and Linkable Resource*, New Jersey et al., 1-22.

Rose, Gillian (1997): "Situating knowledges: positionality, reflexivities and other tactics", in: *Progress in human geography* 21, 305–320.

Rose, Gillian (2001): Visual Methodologies an Introduction to the Interpretation of Visual Materials, London/Thousand Oaks, Calif.

Rottenburg, Richard (2009): *Far-Fetched Facts: A Parable of Development Aid*, Inside Technology, Cambridge, Mass.

Rowley, Jennifer (2007): "The Wisdom Hierarchy: Representations of the DIKW Hierarchy", in: *Journal of Information Science* 33 (2), 163-180.

Sayer, Andrew (2000): For postdisciplinary studies: Sociology and the curse of disciplinary parochialism/imperialism. For sociology: legacies and prospects 83–92.

Schäfer, Mike S. et al. (2015): *Wissenschaftskommunikation im Wandel*, Cologne.

Schellnhuber, Hans Joachim (1998): "Part 1. Earth System Analysis – The Concept", in: Hans-Joachim Schellnhuber/Volker Wenzel (Eds.), *Earth System Analysis: Integrating Science for Sustainability*, Berlin/Heidelberg.

Schellnhuber, Hans-Joachim/Wenzel, Volker (1998): *Earth System Analysis: Integrating Science for Sustainability*, Berlin/Heidelberg.

Schellnhuber, Hans-Joachim/Kropp, Jürgen P. (1998): "Geocybernetics: Controlling a Complex Dynamical System Under Uncertainty", in: *Naturwissenschaften* 85 (9), 411-425.

Schneider, Birgit (2012): "Climate Model Simulation Visualization from a Visual Studies Perspective", in: *Wiley Interdisciplinary Reviews: Climate Change* 3 (2), 185-193.

Schneider, Birgit (2017): "The Future Face of the Earth: The Visual Semantics of the Future in the Climate Change Imagery of the IPCC", in: Matthias Heymann et al. (Eds.), *Cultures of Prediction in Atmospheric and Climate Science: Epistemic and Cultural Shifts in Computer-Based Modelling and Simulation*, 1st edition, Abingdon/New York.

Schneider, Birgit/Nocke, Thomas (2014): Image Politics of Climate Change Visualizations, Imaginations, Documentations, Bielefeld.

Schubert, Cornelius (2015): "Situating Technological and Societal Futures. Pragmatist Engagements with Computer Simulations and Social Dynamics", in: *Technology in Society* 40 (February), 4-13.

Selwyn, Neil (2012): Education in a Digital World: Global Perspectives on Technology and Education. Routledge.

Sengupta, I. N. (2009): "Bibliometrics, Informetrics, Scientometrics and Librametrics: An Overview", in: *Libri* 42 (2), 75-98.

Serway, Raymond A. (1996): *Physics for Scientists & Engineers*. Philadelphia.

Shackley, Simon/Skodvin, Tora (1995): "IPCC Gazing and the Interpretative Social Sciences", in: *Global Environmental Change* 5 (3), 175-180.

Sheppard, Stephen R. J. (2012): Visualizing Climate Change: A Guide to Visual Communication of Climate Change and Developing Local Solutions, Abingdon/New York.

Sismondo, Sergio (1999): "Models, Simulations, and Their Objects", in: *Science in Context* 12 (2), 247-260.

Spieker, Paul Emanuel (1894): "Die Königlichen Observatorien für Astrophysik, Meteorologie und Geodäsie bei Potsdam", in: *Zeitschrift für Bauwesen* 1–3 (46), 1-16.

Star, Susan Leigh (1999): "The Ethnography of Infrastructure", in: *American Behavioral Scientist* 43 (3), 377-391.

Star, Susan Leigh/Bowker, Geoffrey C. (2006): "How to Infrastructure", in: Leah A. Lievrouw/Sonia Livingstone (Eds.), *Handbook of New Media: Social Shaping and Social Consequences of ICTs*, Newbury Park, Calif, 230-245.

Star, Susan Leigh/Griesemer, James R. (1989) "Institutional Ecology, 'Translations' and Boundary Objects: Amateurs and Professionals in Berkeley's Museum of Vertebrate Zoology, 1907-39", in: *Social Studies of Science* 19 (3), 387-420.

Star, Susan Leigh/Ruhleder, Karen (1996): "Steps Toward an Ecology of Infrastructure: Design and Access for Large Information Spaces", in: *Information Systems Research* 7 (1), 111-134.

Steffen, Will et al. (2011): "The Anthropocene: Conceptual and Historical Perspectives", in: *Philosophical Transactions of the Royal Society A: Mathematical, Physical and Engineering Sciences* 369 (1938), 842-867.

Storch, Hans von/Krauß, Werner (2013): Die Klimafalle: Die gefährliche Nähe von Politik und Klimaforschung, München.

Strauss, Anselm (1978): "A Social World Perspective", in: *Studies in Symbolic Interaction* 1, 119-128.

Strauss, Benjamin H. et al. (2015): "Carbon Choices Determine US Cities Committed to Futures below Sea Level", in: *Proceedings of the National Academy of Sciences* 112 (44), 13508-13513.

Sturgis, Patrick/Allum, Nick (2004): "Science in Society: Re-Evaluating the Deficit Model of Public Attitudes", in: *Public Understanding of Science* 13 (1), 55-74.

Suchman, Lucy A. (1988): "Representing Practice in Cognitive Science", in: *Human Studies* 11 (2), 305-325.

Sundberg, Mikaela (2007): "Parameterizations as Boundary Objects on the Climate Arena", in: *Social Studies of Science* 37 (3), 473-488.

Sundberg, Mikaela (2008): "The Everyday World of Simulation Modeling: The Development of Parameterizations in Meteorology", in: *Science, Technology & Human Values* 34 (2), 162-181.

Suneetha, K. R./Krishnamoorthi, R. (2009) "Identifying User Behavior by Analyzing Web Server Access Log File", in: *Computer Science* 7.

Sutherland, Thomas (2018): "Mapping the Space of Flows", in: Sybille Lammes et al. (Eds.), *Time for Mapping: Cartographic Temporalities*, 1st edition, Manchester, UK, 175-196.

Thielmann, Tristan et al. (2012): "Dwelling in the Web: Towards a Googlization of Space", in: *HIIG Discussion Paper Series* 2012-03.

Thomas, William Isaac/Thomas, Dorothy Swaine (1928): *The Child in America; Behavior Problems and Programs*, New York.

Thrift, Nigel (2006) "Re-Inventing Invention: New Tendencies in Capitalist Commodification", in: *Economy and Society* 35 (2), 279-306.

Turkle, Sherry (2009): *Simulation and Its Discontents*, Cambridge, Mass.

Turkle, Sherry (2011): *Evocative Objects: Things We Think With*, Reprint edition, Cambridge, Mass./London.

United Nations (1992): *United Nations Framework Convention on Climate Change*, New York. https://unfccc.int/process-and-meetings/the-convention/what-is-the-united-nations-framework-convention-on-climate-change

van Vuuren, D.P., Kriegler, E., O'Neill, B.C., Ebi, K.L., Riahi, K., Carter, T.R., Edmonds, J., Hallegatte, S., Kram, T., Mathur, R., Winkler, H.

(2014): "A new scenario framework for Climate Change Research: scenario matrix architecture", *Climatic Change* 122, 373–386. https://doi.org/10.1007/s10584-013-0906-1

van Vuuren, D.P., Edmonds, J., Kainuma, M., Riahi, K., Thomson, A., Hibbard, K., Hurtt, G.C., Kram, T., Krey, V., Lamarque, J.-F., Masui, T., Meinshausen, M., Nakicenovic, N., Smith, S.J., Rose, S.K. (2011): "The representative concentration pathways: an overview", in: *Climatic Change* 109, 5–31. https://doi.org/10.1007/s10584-011-0148-z

VanderPlas, Jake (2016): *A Whirlwind Tour of Python*. https://jakevdp.github.io/WhirlwindTourOfPython/, 03.04.2019.

Vaughan, Catherine/Dessai, Suraje (2014): "Climate Services for Society: Origins, Institutional Arrangements, and Design Elements for an Evaluation Framework: Climate Services for Society", in: *Wiley Interdisciplinary Reviews: Climate Change* 5 (5), 587-603. https://doi.org/10.1002/wcc.290.

Verein Lange Nacht der Wissenschaften e. V. (2017): "Lange Nacht Der Wissenschaften 2017", brochure, Berlin. https://www.langenacht-derwissenschaften.de/fileadmin/media/Programm-Archiv/LNDW_Programmuebersicht_2017.pdf

Vertesi, Janet (2014): "Seamful Spaces: Heterogeneous Infrastructures in Interaction", in: *Science, Technology, & Human Values* 39 (2), 264-284.

Von Hippel, Eric/von Krogh, Georg (2003): "Open Source Software and the 'Private-Collective' Innovation Model: Issues for Organization Science", in: *Organization Science* 14 (2), 208-223.

Walsh, Lynda (2014): "'Tricks,' Hockey Sticks, and the Myth of Natural Inscription: How the Visual Rhetoric of Climategate Conflated Climate with Character", in: Birgit Schneider/Thomas Nocke (Eds.), *Image Politics of Climate Change Visualizations, Imaginations, Documentations*, Bielefeld, 81–104.

Wechsung, Frank/Schellnhuber, Hans J. (2018): "Wenn Klimaprojektionen heimkehren: Ungewöhnliche Einsichten für deutsche Flüsse aus einer Klimawirkungsstudie für Pekings Guanting-Region", in: *Potsdam Institute Kurze Mitteilung*, Potsdam.

Wechsung, Frank, et al. (2011): *Elbe-Atlas des globalen Wandels.* Schweizerbart'sche, E., Stuttgard.

Wechsung, Frank et al. (2014): Die Elbe im globalen Wandel: Eine integrative Betrachtung, Berlin.

Wechsung, Frank/Wechsung, Maximilian (2015): "Dryer Years and Brighter Sky – the Predictable Simulation Outcomes for Germany's Warmer Climate from the Weather Resampling Model STARS", in: *International Journal of Climatology* 35 (12), 3691-3700.

Wechsung, Frank/Wechsung, Maximilian (2016): "A Methodological Critique on Using Temperature-Conditioned Resampling for Climate Projections as in the Paper of Gerstengarbe et al. (2013) Winter Storm- and Summer Thunderstorm-Related Loss Events in Theoretical and Applied Climatology (TAC)", in: *Theoretical and Applied Climatology* 126 (3-4), 611-615.

Weinberger, David (2014): Too Big to Know: Rethinking Knowledge Now That the Facts Aren't the Facts, Experts Are Everywhere, and the Smartest Person in the Room Is the Room, First Trade Paper, New York.

Wilderotter, Hans (2005): "Einleitung", in: Hans Wilderotter et al. (Eds.), *Ein Turm für Albert Einstein: Potsdam, das Licht und die Erforschung des Himmels*, 1st edition, Hamburg.

Wilderotter, Hans et al. (2005): Ein Turm für Albert Einstein: Potsdam, das Licht und die Erforschung des Himmels, Hamburg.

Wilkinson, Mark D. et al. (2016): "The FAIR Guiding Principles for Scientific Data Management and Stewardship", in: *Scientific Data* 3 (March), 160018.

Winsberg, Eric B. (2010): Science in the Age of Computer Simulation, Chicago.

Wood, Denis/Fels, John (2008): "The Natures of Maps: Cartographic Constructions of the Natural World", in: *Cartographica: The International*

Journal for Geographic Information and Geovisualization 43 (3), 189-202. https://doi.org/10.3138/carto.43.3.189.

Wyatt, Sally (2005): "Non-Users Also Matter: The Construction of Users and Non-Users of the Internet", in: Nelly Oudshoorn/Trevor Pinch (Eds.), *How Users Matter: The Co-Construction of Users and Technology*, Cambridge, Mass., 67-79.

Yearley, Steven (2009): "Sociology and Climate Change after Kyoto: What Roles for Social Science in Understanding Climate Change?", in: *Current Sociology* 57 (3), 389-40

Cultural Studies

Gabriele Klein
Pina Bausch's Dance Theater
Company, Artistic Practices and Reception

2020, 440 p., pb., col. ill.
29,99 € (DE), 978-3-8376-5055-6
E-Book:
PDF: 29,99 € (DE), ISBN 978-3-8394-5055-0

Elisa Ganivet
Border Wall Aesthetics
Artworks in Border Spaces

2019, 250 p., hardcover, ill.
79,99 € (DE), 978-3-8376-4777-8
E-Book:
PDF: 79,99 € (DE), ISBN 978-3-8394-4777-2

Nina Käsehage (ed.)
Religious Fundamentalism
in the Age of Pandemic

April 2021, 278 p., pb., col. ill.
37,00 € (DE), 978-3-8376-5485-1
E-Book: available as free open access publication
PDF: ISBN 978-3-8394-5485-5

All print, e-book and open access versions of the titles in our list are available in our online shop www.transcript-publishing.com